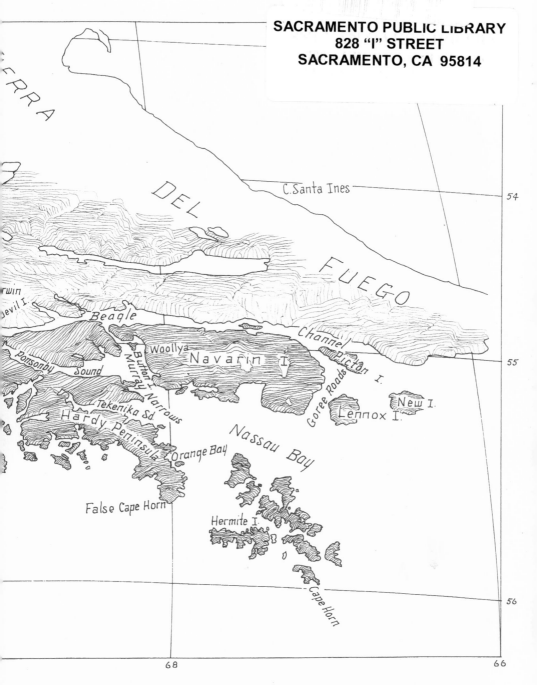

EVOLUTION'S
CAPTAIN

EVOLUTION'S CAPTAIN

The Dark Fate
of the Man
Who Sailed
Charles Darwin
Around
the World

PETER NICHOLS

HarperCollins*Publishers*

HarperCollins books may be purchased for educational, business, or sales promotional use. For information, please write: Special Markets Department, HarperCollins Publishers Inc., 10 East 53rd Street, New York, NY 10022.

Extract from *The French Lieutenant's Woman* by John Fowles, © 1969 by John Fowles. By permission of Little, Brown and Company, Inc.

All maps by Samuel F. Manning © 2003

FIRST EDITION

Designed by Nancy B. Field

Printed on acid-free paper

Library of Congress Cataloging-in-Publication Data is available upon request.

ISBN 0-06-008877-X

03 04 05 06 07 ❖/RRD 10 9 8 7 6 5 4 3 2 1

For Roberta
my ark, my evolution

*But of the fruit of the tree which is in the midst of
the garden, God hath said, Ye shall not eat of it,
neither shall ye touch it, lest ye die.*

*And the serpent said unto the woman,
Ye shall not surely die:*

*For God doth know that in the day ye eat thereof,
then your eyes shall be opened;
and ye shall be as gods . . .*

—GENESIS, 3: 3–5

The TRUE STORY

of the Theft of a Whaleboat

AND

the Subsequent ABDUCTION of
FOUR SAVAGES *from* *Tierra del Fuego*

• • •

Whose *Shocking Misbehaviour* at the
WALTHAMSTOW INFANTS SCHOOL

Led to

CHARLES DARWIN'S
Voyage Aboard the Beagle

• • •

the **Shattering** of
• *Man's Profoundest Beliefs* •

AND

the *Most Ironic* and *Melancholy Fate*

 of its **Captain**
ROBERT FITZROY

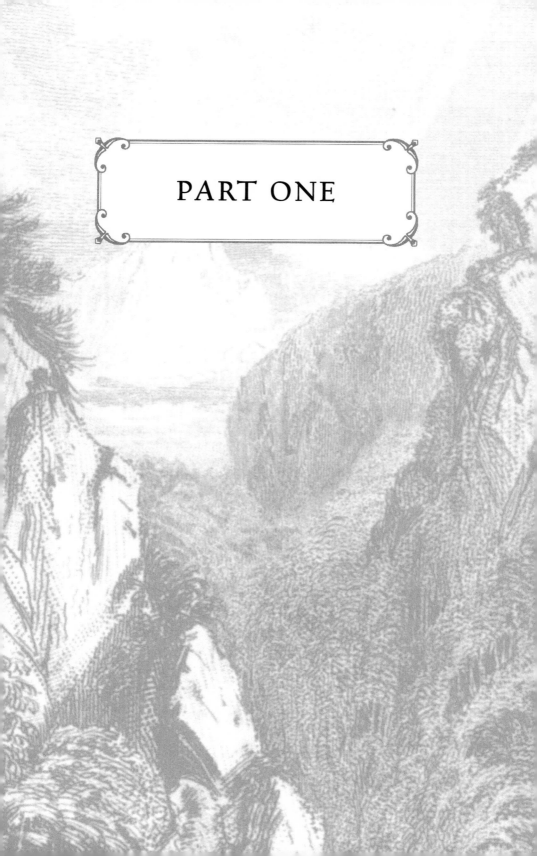

PART ONE

I

Port Famine, Strait of Magellan, August 2, 1828. It is midwinter at the bottom of the world. Snow drives at gale force across the small vessel at anchor. Daylight comes as a few gloomy hours of crepuscular dimness, and the afternoon is already growing dark. Four years later in this same anchorage, in this same vessel even, a young man of unusually sunny temperament—the twenty-four-year-old Charles Darwin—will write in his journal: "I never saw a more cheerless prospect; the dusky woods, piebald with snow, were only indistinctly to be seen through an atmosphere composed of two thirds rain & one of fog; the rest, as an Irishman would say, was very cold unpleasant air."

Alone in his cabin beneath the poop, the vessel's commander, a man still in his twenties, is in the last stage of despair. For him time has lost its swift flow; it has flattened into an unending, intolerable stasis. He sees no relief. He has been in these desolate waters for two years: years more stretch ahead. Home—England, a place as distant as Earth from this cold Pluto—is beyond imagining, beyond regaining.

He raises to his head a small pocket pistol. He is certain of this now, eager for it, and his finger at last tugs with resolve on the trigger.

But there is still too much time: in the long second that

stretches between the release of the hammer, the spark of flint, the flash of powder, and the explosion that sends the ball on its path, his hand wavers, crucially.

He was Captain Pringle Stokes; the vessel, HMS *Beagle*. It lay at anchor in Port Famine with a larger ship, HMS *Adventure*. The two ships, under the overall command of Captain Phillip Parker King, had been dispatched by the British Admiralty in May 1826 to survey the southern coasts of South America, from Montevideo on the Atlantic to Chiloé Island in the Pacific. They were particularly instructed to map what they could of the still largely unknown seacoast of Tierra del Fuego, the desolate, tortuously labyrinthine southernmost tip of the drowned Andes.

The first passage from the Atlantic to the Pacific had been discovered by the Portuguese navigator Ferdinand Magellan, in 1520. He was looking, as was Columbus, as were they all, for that still elusive western route to the spice islands of the Indies. Columbus died in 1506, never knowing he had not found them. It was the Spaniard Vasco Núñez de Balboa who, on September 26, 1513, scaled a hilltop on the isthmus of Darien, in what is now Panama, and first saw the South Sea stretching away in limitless distance beyond Columbus's mistaken Orient. This information expanded the known circumference of the world by more than a third. Seven years later, Magellan, seeking access to that South Sea, found a wide, navigable passage between the bottom of the Americas and, below that to the south, a bleak Terra Incognita. His chronicler, Antonio Pigafetta, who sailed with him, recorded the discovery with an exultant pride:

> We found by a miracle, a strait which we call the strait of the Eleven Thousand Virgins; this strait is a hundred and ten leagues long which are four hundred and forty miles, and almost as wide as less than half a league, and it issues into another sea which is called the Peaceful Sea; it is surrounded by

very great and high mountains covered with snow. I think there is not in the world a more beautiful country, or a better strait than this one.

Magellan's strait is actually 310 miles long from Atlantic to Pacific; but in the weeks they took to pass through it, Magellan and the four ships in his small fleet probably sailed five times that distance. To port, to the south as they tacked endlessly against westerly winds, they saw signs of natives in the dim fires and smoke on the shores of Terra Incognita. Much later, back in Spain, in accounts of the voyage, the land on this southern shore of the strait became known as Tierra del Fuego, Land of Fire.

The only other route from the Atlantic into the Pacific, the open sea passage around a rock mistaken for the southerly cape of Tierra del Fuego, discovered in 1616 by the Dutch captain Willem Schouten of Hoorn, was an exposed and awful place. There, icy winds blast at hurricane force down the glaciers of the Andes, and freak waves driven by westerly winds unimpeded around the bottom of the globe meet off Schouten's false cape in a nightmare maelstrom that was and remains a desperate place for any vessel. Seeking a fast passage to Tahiti in 1788, William Bligh tried to force his ship, HMS *Bounty*, past this Cape Horn. He spent over a month tacking back and forth, making only a handful of miles to westward in all that time. He, and more especially his crew, became so demoralized that he turned around and sailed the other way to Tahiti, eastward around a good part of the world, just to have the winds at his back.

Bligh knew too little about the twisting Strait of Magellan immediately to his north to force his ship through it with a fractious crew. Forty years later, blizzard-bound on its northern shore, the *Adventure* and the *Beagle* were attempting to chart a safe passage through the strait, to find a less forbidding route for ships passing between the Atlantic and the Pacific.

*   *   *

Though neither their commanders nor the lords of the Admiralty who had penned their orders saw them as such, the two ships anchored in Port Famine were part of the grandest design of history—so audacious that not until it was in place was it truly seen by those who had made it. In this year of 1828, the British East India Company had been flourishing for two centuries; the Raj, a deeply entrenched and structured community of 50,000 British soldiers, merchants, and their families, controlled 90 million inhabitants in India, the biggest, proudest, most ostentatious foreign possession of the suddenly *Great* Britain. But even before the chubby eighteen-year-old Victoria acceded to the throne in 1837, British interests around the world were acquiring the solidity and permanence of the railway network and its massive gothic stations that were being built at home.

Simple geography suggested what was soon to happen: In the sugar islands of the Caribbean, in the Canadian Arctic, at the Cape of Good Hope, in Singapore and Penang and Bermuda, in British Honduras and British Guiana, in Gibraltar and the Falkland Islands, Mauritius, Ceylon, Hong Kong, Australia and New Zealand, on the coasts of East and West Africa, and pushing into its continental heart, the British were taking control, subjugating the natives, settling, sowing, mining, and taking away whatever could be used at home or traded for something elsewhere. They brought God and moral certainty with them, and a rigidly hierarchical society into which, in servile positions, the locals everywhere fit perfectly. They had staffed these outposts, or "stations"—little Englands landscaped, groomed, and tidied to resemble home—with over a million Britons. They were defended, and communication between them was ensured, by the armies and navies that had so recently defeated Napoleon and become the greatest war machine on Earth. All these far-flung trading stations were, for a time, simply British "interests" abroad.

But they could not remain so. Power does not hover at a flat-line: It must be protected and fortified; it must grow, exponentially, or collapse. There could never be enough tea, diamonds, ivory, gold, cotton, and lumber; eventually mere supremacy and influence would no longer suffice. With the Industrial Revolution's starburst of refinement in the sciences of steam and ironmaking, with improvements in transportation and mining and manufacturing and the building of tunnels and bridges and railways, alloyed with the inexhaustible abundance of cheap native labor, Britain's engineers, merchants, and emissaries of government swarmed over the earth, protected and boosted by the might of British power and moral sovereignty. And long before anyone imagined or wanted such a thing, the little island nation found itself possessed of an empire, on a fabulous and unprecedented scale, and it seemed right and natural. Back home in England, mapmakers and printers of atlases agreed on a certain pastel pink to delineate these possessions, and Englishmen spinning their globes in paneled clubrooms could follow beneath the glossy lacquer a swath of pink geography so completely encircling the earth that, one saw immediately, the sun never set on the British Empire.

A sea passage through the Americas, between east and west, was obviously an integral and necessary part of such a design. (A canal cut through the mountainous isthmus of Central America had been suggested by a priest, Francisco López de Gómera, as long ago as 1552, but three centuries later this was still a remote possibility.) The few existing charts of the southern coasts, made by Magellan, Sarmiento, Schouten, and others, were inaccurate, inadequate, and worrisome: others would come wanting to fill in the blanks. In the early nineteenth century, the continent of South America tantalizingly resembled a second Africa, but it was far less known and still largely unpenetrated. The wealth of its minerals and resources—El Dorado and all that Inca gold that had inflamed the fancies of the Spanish conquistadores—could still only be guessed at. The influence of Spain had dwindled, leaving the way open for the courtship of Latin America's nascent coastal states by France and the newly autonomous

North American union. Apart from the Arctic and the Antarctic, which had only been distantly seen and guessed at by explorers such as Cook, the bottom of South America was as unknown as any place on Earth—remoter and more unexplored by Europeans than Africa or Borneo. As a gateway between the oceans, and a site of strategic importance, this continental extremity was unique. An accurate charting of the South American coasts was imperative. This would reveal harbors of refuge, military potential, and any alternative passages to Cape Horn and the Strait of Magellan. It would mean control and influence over more, history instructed, than could be imagined.

The two ships made their headquarters at Port Famine, roughly halfway through Magellan's strait. The port's name describes the fate of the harbor's first European inhabitants. They were put ashore in 1584 by the Spanish navigator Pedro Sarmiento, in an attempt to establish a Spanish colony, which Sarmiento named San Felipe, after his king. In the summer of 1584, it seemed a promising place: then as now a natural harbor, easy to sail in and out of, well protected from prevailing winds, with a good range of depth for all sorts of vessels, a bottom that provided good holding for anchors in clay and sand. There were convenient landing places ashore, and thickly wooded hillsides offered timber for fuel and building. But when winter came, Sarmiento's settlers starved and froze. The few survivors were rescued by a passing English ship, whose captain, Thomas Cavendish, gave the place a name that stuck. Today part of Chile, it is called Puerto del Hambre, Port of Hunger.

Captain Phillip Parker King was aware of the place's grim history, but the *Adventure* and the *Beagle* were well stocked with food and equipment, and he chose to make Port Famine, with all the advantages it offered, the base for their operations. He split the territory to be surveyed in two. He took *Adventure* east and south and sent Captain Pringle Stokes and the *Beagle* west, up the narrowing strait toward the wilder, more labyrinthine Pacific shore, into the area of greater blanks on their maps. Stokes sailed, literally, off the charts.

Our twenty-first century satellite-corrected atlases reveal the land here as a maze of mountain-sized, long-limbed Rorschach blots, between and through which course sizable crumbs of glacier ice and seawater driven by storms and ferocious tidal streams. Today's sailors navigate these waters with satellite-enabled handheld GPS receivers. If they are exposed on deck, their heads are wrapped in fleece hats, tucked inside Gore-Tex hoods. Eyes dart between digital readouts and views of cloud-hung sounds and fjords, each one resembling the next. It is almost impossible to imagine how one would proceed without this miraculous technology, without a diesel engine throwing a big-bladed propeller, without central heating and the space-age clothes that keep a sailor warm and, above all, dry.

Before the global positioning system (GPS) had been created, the few seafarers venturing here sailed chart in hand, marking off each headland, each rock, as one would find a route with a Baedecker through Budapest at night in fog. But even then the weather, obscuring and revealing the poorly glimpsed tumbling, haphazard land, distorting its apparent size and distance, could confuse the most careful navigator. Stokes too had charts, early Spanish editions, and those drawn by the English navigator Sir John Narborough, which were little better than the ancient vellums on which priests had marked the inferences of Hades. The most sophisticated navigational devices of his age, sextant and chronometer, depended on sighting the sun or the moon or the stars, and this, beneath the unceasing roll of storms that came off the Southern Ocean, was rarely possible. In effect, Stokes pushed the tubby square-rigged *Beagle* through fog while negotiating hurricanes.

For three months, from February to April 1827, the *Beagle*

explored the western reaches of the Strait of Magellan. Charting the serrated coast in an unbroken line was an impossible ambition, not even attempted. (A modern chart, corrected to the year 2001, shows large areas of coastline in the western part of the strait marked with dotted lines, indicating even now the uncertainty of contour. It bears the written warning: "Owing to the incomplete nature of the surveys this chart should be used with caution.") What Stokes did attempt was the accurate plotting of points crucial to navigators. Within a mile or two of where they are shown on a chart today, he fixed the latitude and longitude of Cape Victory, actually a narrow islet, and Cape Pillar, the tip of Desolation Island, which together define the western entry to the Strait of Magellan, and between them a visible but hazardous scattering of rocks called, by the Spanish, the Evangelists. The fixing of these points, and a few bays and anchorages, with comments on what might be of use in them to passing ships, were the results of these months of perilous navigation. Stokes's main accomplishment each day was preventing the loss of his ship. This was a constant danger, and many times appeared likely, and the toll on the man responsible was deeper than even he knew at the time. Finally, in April, he turned his ship downwind at last and headed back to rendezvous with the *Adventure* in Port Famine. From there the two ships sailed north to Montevideo and Rio de Janeiro, where they spent the southern winter and replenished their stores.

When they returned for a second season of surveying, Stokes was instructed to proceed again to the western part of the strait, and from there continue surveying north as far as Chiloé Island. He sailed once more from Port Famine in March 1828, eager, as far as his commanding officer Captain King knew, to add to his work of the year before. But this second plunge turned his mind. Into his plainly written journal, a normally staid, even dull accounting of daily navigation and notable features passed on the coast, there crept a maggot of dissonance, the first tremors of a sickness that would ineluctably waste and finally destroy:

Near Cape Notch the mountains spire up into peaks of great height, singularly serrated, and connected by barren ridges. About their bases there are generally some green patches of jungle; but, upon the whole, nothing can be more sterile and repulsive than the view.

The scenery in the Strait of Magellan should not have looked repulsive or otherworldly to Stokes. In places it resembles the European Alps, or the magnificence of the Scottish Highlands, which had been widely appreciated in the romantic fiction of Sir Walter Scott. (Stokes's use of the word "jungle" to describe vegetation is misleading: what he saw was a dense scrim of tundra mosses and thigh-high forests of storm-bent trees; average temperatures ashore hovered year-round at just above freezing; snow frequently settled over this "jungle.") In the more protected fjords, the landscapes resemble those described by Vancouver and Cook in their accounts of travels along the west coast of North America and Australia and New Zealand, or the early nineteenth-century paintings of the Hudson River Valley and the untrammeled spaces of the new United States, all of which had set people in England to talking about the sublimity of such natural edens.

But Stokes would not see it. He did not agree with Pigafetta that "there is not in the world a more beautiful country." To him it became a malignant vision. He began to hate it with a poison that worked its way through him, through his journal, and it appeared inescapable:

The coast between Capes Isabel and Santa Lucia is dangerous to approach nearer than ten miles, for there are within that distance many sunken rocks . . . the general aspect of this portion of the coast is similar to the most dreary parts of the Magalhaenic regions.

The conditions through which he attempted to push the *Beagle* did nothing to lighten his view.

By 8 pm we were reduced to the close-reefed main-topsail and reefed foresail. The gale continued with unabated violence during the 6th, 7th, and 8th (April), from the north, N.W. and S.W., with a confused mountainous sea. Our decks were constantly flooded . . . the little boat which we carried astern was washed away by a heavy sea that broke over us . . . the marine barometer was broken by the violent motion of the vessel. . . .

The effect of this wet and miserable weather, of which we had had so much since leaving Port Famine, was too manifest by the state of the sick list, on which were now many patients with catarrhal, pulmonary, and rheumatic complaints . . .

The ship was frequently stormbound, frustrating Stokes's attempts at progress and the success of his mission.

Nothing could be worse than the weather we had during nine days' stay here [Port Santa Barbara]; the wind, in whatever quarter it stood, brought thick heavy clouds, which precipitated themselves in torrents, or in drizzling rain . . .

Ship management was always difficult. At times boats had to be lowered to drag the *Beagle* from possible shipwreck.

After running two miles through a labyrinth of rocks and kelp, we were compelled to haul out, and in doing so scarcely weathered, by a ship's length, the outer islet. Deeming it useless to expend further time in the examination of this dangerous portion of the gulf, we proceeded towards Cape Tres Montes . . .

Stokes pushed on up the western shores of Patagonia for another two months. His journal revealed a mounting catalog of torment and failure, and a pathological estrangement from his work.

Exceedingly bad weather detained us at this anchorage. From the time of our arrival on the evening of the 21st, until midnight of the 22d, it rained in torrents, without the intermission of a single minute, the wind being strong and squally at W., W.N.W., and N.W. . . .

Another day and night of incessant rain. In the morning of the 25th we had some showers of hail, and at daylight found that a crust of ice, about the thickness of a dollar, had been formed in all parts of the harbour. . . .

Here we were detained until the 10th of June by the worst weather I ever experienced. . . . Nothing could be more dreary than the scene around us. The lofty, bleak, and barren heights that surround the inhospitable shores of this inlet, were covered, even low down their sides, with dense clouds, upon which the fierce squalls that assailed us beat, without causing any change. . . . Around us, and some of them distant no more than two-thirds of a cable's length, were rocky islets, lashed by a tremendous surf; and, as if to complete the dreariness and utter desolation of the scene, even the birds seemed to shun its neighbourhood. The weather was that in which . . . "the soul of a man dies in him."

It got worse. The ship's 28-foot-long yawl, "a beautiful boat," and vital to their surveying methods, was smashed to pieces while being hoisted aboard in a gale: "We were obliged to cut her adrift. . . . her loss was second only to that of the ship."

Stokes's crew, despite being outfitted with foul weather "clothing"—lengths of painted canvas to wrap around themselves (this only ensured a clammy discomfort)—began to fall apart dramatically. They were generally hardy young men, in their teens through their thirties, but without being able to get ashore and supplement the ship's salt beef and pork and rock-hard biscuit with fresh meat, they soon began to suffer from scurvy. Their gums bled and their teeth loosened, old scars opened, they grew listless and weak, and the cold accelerated this frighteningly, until it seemed that Stokes's crew might not be able to control the ship. After consulting with his surgeon, Mr. Bynoe, Stokes made for a landlocked anchorage where the *Beagle* was temporarily decommissioned for a period of convalescence. The yards and topmasts were struck, the ship was covered with sails for protection, the crew were put on light duty, the sick were tended. Even safe from storms, their harbor of refuge . . .

> being destitute of inhabitants, is without that source of recreation, which intercourse with any people, however uncivilized, would afford a ship's company after a laborious and disagreeable cruise in these dreary solitudes.

And here Stokes's journal stops. Even he must have tired of the repetition of his lamentations.

After a two-week rest, the crew somewhat revived, the *Beagle* sailed south and east heading back to the Strait of Magellan and Port Famine, where the *Adventure* was waiting for her. The passage took almost four weeks and Captain Stokes remained in his cabin the entire time. The ship was effectively commanded by his assistant surveyor, Lieutenant William Skyring, and the ship's sailing master, Samuel Flinn.

Fighting strong winds to the last, even tacking into sheltered water, the *Beagle* reached Port Famine after dark on the wintry evening of July 27. Skyring immediately had the bosun's gig row him across to the *Adventure* where he climbed aboard and reported Stokes's condition to Captain King. King went to see for himself.

"I went on board the *Beagle* in the evening and found Captain Stokes, after the first two or three minutes, perfectly collected and communicative of all the events of his cruize," he later wrote in a long letter to the Admiralty. "For three days afterwards I saw him daily during which he resumed all his duties with increased energy."

However, King told his ship's surgeon to confer with Benjamin Bynoe, the *Beagle*'s doctor, so both men could give him an opinion of Stokes's health. They told him that although Stokes had at times in the last few months "expressed himself weary of life and wished to meet his death," he now seemed so recovered that they thought he might be able to carry on his duties. Those duties meant continuing the surveying mission for several more years at least. While the two surgeons were aboard the *Adventure* actually reporting this to King, word came from the *Beagle* that Stokes had shot himself.

They found him in his cabin, his head streaming blood, his linen shirt and clothes drenched, blood slippery on the cabin sole, but Stokes was quite conscious and even apologetic for making himself a nuisance. The two doctors immediately did what they could for him, which wasn't much: a head wound, no exit wound, the pistol's ball lodged somewhere inside. They cleaned him up and laid him in his bunk.

For the next three days, King spent hours aboard the *Beagle* with Stokes, who remained "perfectly collected." As they talked, King went through Stokes's papers, his journal and surveying observations, the ship's daily memoranda, accounts, and orders. These "were in so confused and scattered a condition," wrote King, "that I despaired of putting them into any order."

Stokes was overwhelmed with remorse. He told King that the main reason for his "unhappy malady" was his fear that he wasn't up to the job, that his defects as a surveyor were bound to come to light. And they did, as King began to see that most of what had been accomplished on the *Beagle*'s cruise—the charts, the harbor plans, the laborious azimuths and bearings and calculations made in the course of surveying—was the work of Lieutenant Skyring and the junior officers. The calculations written in Stokes's hand were actually copies of what had been done by the others. This was never admitted in King's subsequently published account of the cruises of the *Beagle* and the *Adventure*, in which he praised Captain Stokes for his work and stoicism. In his letter to the Admiralty, however, King shifted his praise to the ship's master, Flinn, for extricating the *Beagle* from "situations of impending danger into which her Commander had unwarily and rashly rushed without any regard to the lives of so many people under his protection. . . . The state of Captain Stokes's mind drove him at times to such desperate acts, as regarded the conduct of the ship, in which he would be controlled by no one, but (when the case arrived at a pitch of extreme danger) by the Master Flinn." Before returning to Port Famine, Stokes had extracted a pledge from his officers that they would never tell what had happened. But now, lying on his bunk, his deception revealed, he told King everything and praised Flinn with almost exactly the same self-recriminatory words.

At moments, as he lay in his bloody berth, Stokes even talked of resuming command once he recovered, as he began to feel he would. But after three days of excited chatter and confession he worsened. Gangrene slowly made its way through his brain. It took him twelve days to die. The first entry in the *Beagle*'s log for August 12, 1828, made just after midnight, reads: "Light breezes and cloudy. Departed this life Pringle Stokes, Esq., Commander."

After death, his body was examined. Despite his tremorous hand, the gunshot had done its job. The surgeons, duty bound to provide postmortem evidence, opened Stokes's head and found the pistol's small-bore ball lodged in the corrupted mess of his brain.

They also found seven nearly healed knife wounds in Stokes's chest: the inept captain had been trying to kill himself for weeks.

The two ships then sailed north for the winter. The crews of both vessels were weak from months of exposure and hard service on insufficient rations. They had caught fish and shot what game they could find, but it was never enough, and they were plagued by scurvy. At Montevideo they took aboard a supply of bitter Seville oranges and these alone had every man better in less than a week.

The *Beagle* spent six weeks in Montevideo undergoing repairs to its hull, while King sailed north in the *Adventure* to Rio de Janeiro, where he was to report to Sir Robert Otway, the commander-in-chief of the South American fleet, aboard the fleet's flagship, HMS *Ganges*. After two seasons in the remote south, much of the surveying work commissioned by the lords of the Admiralty was still unaccomplished. Despite the hardship that had driven one captain to suicide, killed several officers, and left men on both ships weakened from scurvy, *Adventure* and *Beagle* were to go south again.

On Stokes's death, King had appointed Lieutenant Skyring as the *Beagle's* acting commander. The promotion, when Skyring assumed it on August 12, 1828, was still unofficial; not until confirmed by Sir Robert Otway would he be formally recognized as the *Beagle's* new captain. However, with his accomplishments aboard the ship under conditions harrowing inboard and out, the captaincy appeared to be his.

But Otway had his own ideas. He superseded Skyring with a favorite of his own, the *Ganges'* lieutenant, twenty-three-year-old Robert FitzRoy. No doubt King remonstrated to the extent he felt able with his commanding officer in the privacy of Otway's cabin aboard the *Ganges*. Skyring was the obvious choice: he knew the *Beagle* intimately, he had handled both the ship (along with Master Flinn) and her former captain with delicacy and skill, and although nominally the assistant surveyor, he had done the main part of that work while Stokes was going to pieces in his cabin. Otway was not swayed. Stokes had also been King's choice, and now Otway wanted his man aboard the *Beagle*.

King could only submit to his superior's wish. Years later, in the published account of his South American voyage, he still politely objected:

> Although this arrangement was undoubtedly the prerogative of the Commander-in-chief . . . it seemed hard that Lieutenant Skyring, who had in every way so well earned his promotion, should be deprived of an appointment to which he very naturally considered himself entitled. . . . (Lieutenant) FitzRoy was considered qualified to command the *Beagle*, and although I could not but feel much for the bitterness of Lieutenant Skyring's disappointment, I had no other cause for dissatisfaction.

When the *Beagle* arrived in Rio from Montevideo, Skyring was relieved of command. Robert FitzRoy went aboard at 6 A.M. on December 15, 1828, and assumed his captaincy. Lieutenant Skyring took up his former position as assistant surveyor and served his new captain with loyalty, goodwill, and without bitterness.

In the same way that a commander slain on the battlefield is replaced by an unknown who will profoundly change the outcome of a war, Robert FitzRoy now stepped into the light of his peculiar destiny. He hoped for advancement and distinction, and these he duly achieved. But he would find them in the shadow of a fame that obliterated these accomplishments. He is remembered only for his pivotal role in what he came to consider an abomination.

2

In 1934, Nora Barlow, the granddaughter of Charles Darwin, visited Robert FitzRoy's elderly spinster daughter, Laura FitzRoy, at her home in London. The dark, cluttered Victorian drawing room, resisting any influence of the twentieth century, was dominated by a large white marble bust of Miss FitzRoy's father. "A remarkable face," Nora Barlow later wrote, "sensitive, severe, fanatical; combining a strength of purpose with some weakness or uncertainty."

He was also handsome. Drawings of FitzRoy as a young man show a long, thin, aquiline nose; the smudged impression of a wispy moustache; limpid, long-lashed, feminine eyes; and a high forehead, suggesting that the dark hair combed forward at the top and along the temples covered a receding hairline. It is indeed a sensitive, narrow, introspective, finely featured face, and in the line of the thin lips, in a certain lift in the brow, there is more than a hint of a volatile mix of intellectual brilliance, intolerance, and arrogance. It could be the face of a supercilious young classics scholar at Oxford too sure of his own capacity, or of a young Sherlock Holmes—picture his first Hollywood incarnation, the chilly, long-nosed actor Basil Rathbone, as a twenty-three-year-old. Undoubtedly FitzRoy liked the way he looked and paid attention to his appearance.

Robert FitzRoy as Darwin first knew him, at about the time of his second voyage aboard the *Beagle*. (*Drawing of Robert FitzRoy by Phillip Parker King, by permission of the Alexander Turnbull Library, Wellington, New Zealand*)

He was a great believer in phrenology, the bogus science in such vogue during the nineteenth century, that supposed that the shape and size of the human cranium, together with its telling bumps, determined character and mental faculties. FitzRoy drew instant conclusions about people based on such evidence. He had a distrust of people with coarse features or wide spatulate noses. He believed them untrustworthy and inferior.

In one sense, Robert FitzRoy himself was exactly what he appeared to be: a highly strung aristocrat. The FitzRoys had descended as the dukes of Grafton, a favored though illegitimate branch of royalty, from a liaison between Barbara Villiers and King Charles II. For generations, the Graftons were at the highest level of British society. They were members of the royal court, they were high-ranking Tories in government, they owned great estates in England and Ireland. They legislated and they controlled. At least three dukes of Grafton were admirals, men used to getting their way in every respect. One of them, Lord Augustus FitzRoy, third duke of Grafton, commanded a small fleet of four vessels to fire on four passing French ships in the Caribbean one day in 1737 when they wouldn't stop at his hail, this precipitated a full-scale naval action between all eight ships. Britain and France were not at war at the time; the battle finally ended with gracious apologies on both sides, and the ships sailing on.

Another duke of Grafton was briefly acting prime minister in 1767, between the governments of the ailing William Pitt the Elder and Lord North. FitzRoy's father, Lord Charles FitzRoy, became a general in the British Army, an aide-de-camp to George III, and a member of Parliament. His mother, Lady Frances Anne Stewart, came from an equally dominant strain; she was the eldest daughter of the first Marquis of Londonderry.

FitzRoy's connections were impeccable, except in one respect: from his mother he inherited a kinship with mental instability. In 1820, when Robert FitzRoy was fifteen, his uncle, the third Marquis of Londonderry, who had teetered for years on the edge of madness, finally committed suicide, with terrible efficiency, by slashing his own throat with a razor. His uncle's death and its manner made an impression on the boy that he never forgot. He spoke about it and feared it all his life. It loomed forever just beyond the foreground in his mind as a specter of his own predisposition.

Sir Robert Otway's faith in him was justified: FitzRoy was certainly capable. In his competence and zeal he was light-years from Pringle Stokes.

He had been sent off to the Royal Naval College at Portsmouth when he was only twelve years old—possibly he was already exhibiting an unusual precocity at his studies and was deemed ready for more, but more probably because of the cold practice among the English upper classes of sending small children away to school at a still tender age, breaking off their childhood with a brutal snap. At Portsmouth, in addition to classical studies in Greek, Latin, mathematics (including the spherical trigonometry necessary for celestial navigation), French, drawing and painting, fencing, and dancing, the syllabus included:

> Naval history and nautical discoveries; naval architecture; astronomy, motions of heavenly bodies, tides, lunar irregularities; the *Principles* and other parts of Newton's philosophy; fortifications, doctrine of projectiles and its application to gunnery; principles of flexions and application to the measurement of surfaces and solids; generation of various curves, resistance to moving bodies; mechanics, and hydrostatics.

FitzRoy tore through the three-year course at Portsmouth in less than twenty months and graduated with full marks and a gold medal, neither of which had ever been awarded to any cadet. In 1819, at the age of fourteen, he went to sea. He served aboard ships in the Mediterranean and with the South American fleet, where he became a favorite of Sir Robert Otway. There was an inquiring, scientific bent to his mind, and by the time of the "death vacancy" resulting from Pringle Stokes's suicide, FitzRoy was abreast of the latest thinking in geology, paleontology, and "natural philosophy," as the study of nature was then known. It was as well for him that Otway liked him, and that his family was intimately connected with the Admiralty and the powers that determined advancement. He hardly needed these connections to get ahead, but a man of such dazzling and superior ability, with self-assurance, vanity, willfulness, and a personal fortune, probably made few friends. He was admired but rarely liked. Otway had fought with Nelson at Copenhagen and had been the legendary admiral's flag captain; perhaps he saw something of himself in the ambitious young FitzRoy.

It was a plum assignment. Without the opportunities provided by war, the captaincy of the *Beagle* offered FitzRoy a unique chance to prove himself—more than he could imagine.

FitzRoy and the *Beagle* were kismet for each other.

The young captain's new command was a 90-foot-long, 235-ton, 10-gun brig; a member of the Cherokee class (named for the first to be built), designed in 1807 by Sir Henry Peake, Surveyor of the Navy. Its firepower comprised eight 16- or 18-pound carronades and two 6-pound chase guns. Carronades (named for the Carron Iron Founding and Shipping Company that invented them) were short, light carriage guns that fired heavy shot over a short range. They were developed for close-in brawling in the Napoleonic naval battles of the late eighteenth and early nineteenth century. These small but powerful guns allowed a smaller ship to carry them, thus reducing the size and costs, for a given firepower, of certain classes of fighting vessel for the Admiralty.

In size and armaments, the Cherokee-class ship was about halfway between the *Sophie* and the *Polychrest*, Captain Jack Aubrey's first two commands in Patrick O'Brian's Aubrey/Maturin series of naval adventure novels. They were handy, economical in-fighters, but after the wars were over many went on to peacetime work, and it was discovered that they could be a liability at sea. They had low freeboard, and waves readily broke over their decks; high, solid bulwarks prevented this water from easily running off back into the sea. Several heavy waves breaking over the deck of a Cherokee-class ship could quickly make her top-heavy and, in a stiff breeze, capsize her. Almost a quarter of the approximately 100 Cherokees built were wrecked or foundered in heavy weather. Seamen called them "coffin brigs."

Yet they sailed well, for a square-rigged vessel; they were sea-kindly, with a comfortable motion; deep, capacious holds could carry an enormous quantity of stores and spares, clear decks held numerous boats, gear, equipment—qualities that would be desirable in a modern cruising yacht.

The keel of the forty-fifth vessel to be built to this design was laid in 1818 at Woolwich Dockyard on the River Thames. The ship was launched exactly two years later, at a cost of £7803. She was christened *Beagle*—history has left no record why; perhaps the whim of some hunting admiral. She was placed "in ordinary"—that is, tied up with nothing to do—for five years until she was allocated to the Navy's surveying service in 1825, which commissioned her to sail with HMS *Adventure* to South America. By this time, however, some of her planking was already rotting in the tidal river water, so she was sent back to Woolwich dockyard for repairs and much else that was needed for her coming voyage. The bulwarks were lowered to allow water to drain off more easily; a chartroom was built over the quarterdeck, more cabins built forward of this new poop deck for additional storage and an assistant surveyor; skylight hatches were fitted over the captain's cabin and the gun-room, where the officers slept, to make the long months and years aboard less gloomy. A mizzenmast, carrying two fore-and-aft (rather than square) sails, was added to her stern, greatly increasing her maneuverability. This changed her rig from a two-masted brig to a barque.

Six months after she went into the dockyard, on May 22, 1826, HMS *Beagle* sailed from England on her first mission to

Tierra del Fuego, under the command of her first captain, Pringle Stokes.

Designed for war, one of a bunch built for a price, she sailed away to explore the unknown world, into maelstroms of unholy wind and weather, and she proved herself one of the ablest little ships in history.

3

In January 1829, HMS *Adventure*, with Captain King, HMS *Beagle*, with her new captain, Robert FitzRoy, and a smaller schooner, HMS *Adelaide*, sailed south again on their surveying mission. (Not often mentioned, the *Adelaide*, named for England's queen, the wife of William IV, had been purchased by Captain King earlier in the expedition, and had accompanied the *Adventure* for much of 1828. Being a "fore-and-after" rather than a square-rigged vessel, she was a handier sailer, able to point closer to the wind than either of the other two ships, and was used for survey work in tighter channels and harbors.)

Captain FitzRoy encountered misfortune early in his command. On January 30, off Maldonado on the coast of Uruguay, the *Beagle* was caught by a *pampero*, a vicious squall blowing off the pampas, later said to be the worst in many years. It lasted only twenty minutes, but the *Beagle*, newly loaded to the brim with supplies, exacerbating the already top-heavy tendency of her class, was knocked over on her beam, and lay pinned by blasts of wind for long minutes, during which it appeared she might capsize. Topmasts, topsails, all sorts of sails and small spars, were torn away and blown to pieces. Two seamen who had been furling sail high in the rigging were blown away with them into the sea and lost.

Such sudden "white squalls" can catch a ship by surprise and knock it down in seconds, long before its captain has time to notice what's coming and take in sail. The incident was not FitzRoy's fault (though in later years he was to blame his inexperience for not being more alert and ready for such a possibility), but to lose men so early in his command would have had his crew (seamen are a highly superstitious lot) wondering if their new skipper had the curse of being unlucky.

Two months later, the ships reached the eastern entrance to the Strait of Magellan.

On April 13, as the *Beagle* was beating down into the strait near Cape Negro against a light southerly breeze, the crew spotted some natives: two women and a child in a canoe near the shore, two men and their dogs close by on the beach. FitzRoy had a whaleboat lowered and a crew of seamen pull him shoreward for a closer look, the people "being the first savages I had ever met."

To Englishmen, they appeared remarkably unattractive. "hideous . . . filthy and most disagreeable" was Captain King's first and lasting impression.

To the Fuegians, the British naval officers and their crew, appearing from seaward in their grand and intricate vessels, with their elaborate clothing, their gadgets, and their inexplicable powers, were as otherworldly as little green men coming out of a spaceship—except that the locals were no longer astonished. Such visitors had been turning up for some time; the occasional glimpses of the Magellans, Sarmientos, and Schoutens had increased to nearly regular traffic. By 1827, the year King's expedition reached the Strait of Magellan for the first time, sealing and whaling vessels (usually ignored by history because their crews didn't plant flags, slaughter locals, colonize, or make claims for distant sovereigns) had been passing through Tierra del Fuego, harvesting seals and penguins, and trading with the natives for more than fifty years. Relations had evolved considerably since the earliest contact. The Fuegians had largely forgotten their initial fears and become instead cheeky opportunists. It was the latecomers on the scene, the Englishmen in 1827, who

Fuegian native of the "Yapoo Tekeenika" tribe, as FitzRoy mistakenly believed they called themselves. (Narrative of HMS *Adventure* and *Beagle*, by Robert FitzRoy)

were the naive ones. They were smugly and vastly amused by their own abilities to impress the locals with their beads and cheap magic, underestimating not only the Fuegians' cleverness, but their self-respect.

"They were pleased by a ticking watch," wrote King, of his first encounter with a group of them. While dazzling them with his watch, he surprised a Fuegian by deftly cutting off a lock of the native's hair with a pair of scissors (perhaps he fancied a small trophy brush fashioned of the coarse hair by the ship's carpenter, not an unusual item). The man objected until King gave the hair back to him. "Assuming a grave look, he very carefully wrapped the hair up, and handed it to a woman in the canoe, who, as carefully, stowed it away in a basket . . . the man then turned round, requesting me, very seriously, to put away the scissors, and my compliance restored him to good humor."

At another encounter, "one of the party, who seemed more than half an idiot, spit in my face; but as it was not apparently done angrily, and he was reproved by his companions, his uncourteous conduct was forgiven." The Fuegians spat at them, and the English sneakily cut off their hair; what was courteous or idiotic was misunderstood by both sides.

Two years later, Robert FitzRoy saw them in much the same way: they were dirty and primitive. And he brought his own elevated learning, particularly his ideas about physical appearances and phrenology, to the deduction of their innate character.

Their features were . . . peculiar; and if physiognomy can be trusted, indicated cunning, indolence, passive fortitude, deficient intellect, and want of energy. I observed that the forehead was very small and ill-shaped, the nose was long, narrow between the eyes and wide at the point; and the upper lip, long and protruding. They had small, retreating chins; bad teeth; high cheekbones; small Chinese eyes at an oblique angle with the nose. . . . The head was very small, especially at the top and back; there were very few bumps for a craniologist.

FitzRoy was not quite so scientific about the younger of the two women in the canoe. She "would not have been ill-looking, had she been well-scrubbed, and all the yellow clay with which she was bedaubed, washed away."

It was the late, too tender Pringle Stokes who saw these people most clearly. His journal descriptions of his first encounters with Fuegians in 1827 were full of carefully observed details, largely without the prejudicial filter of a self-righteous European sensibility.

As might be expected from the unkindly climate in which they dwell, the personal appearance of these Indians does not exhibit, either in male or female, any indications of activity or strength. Their average height is five feet five inches; their habit of body is spare; the limbs are badly turned, and deficient in muscle; the hair of their head is black, straight, and coarse; their beards, whiskers, and eyebrows, naturally exceedingly scanty, are carefully plucked out . . . the mouth is large, and the under-lip thick; their teeth are small and regular, but of bad colour. They are of a dirty copper colour; their countenance is dull, and devoid of expression. For protection against the rigours of these inclement regions, their clothing is miserably suited; being only the skin of a seal, or sea-otter, thrown over the shoulders, with the hairy side outward.

They also smeared themselves with seal oil and blubber, which "combined with the filth of their persons, produced," to Captain King, "a most offensive smell." This seemed manifest lowliness, the benighted savage rolling in filth. The Englishmen didn't realize it was an effective weatherproofing, something perfectly suited to the climate. No contemporary clothing, no oiled or painted canvas, could keep the densely wet weather of Tierra del Fuego from reaching the body. The English sailors' habit of wrapping themselves in clammy, moisture-retaining layers ensured constant misery: "Our discomfort in an open boat was very great, since we were all constantly wet to the skin," they complained.

Stokes carefully noted the natives' diet—shellfish, seal, sea-otter, porpoise and whale, wild berries, and certain seaweeds—and the fact that they weren't particular.

> Former voyagers have noted the avidity with which they swallowed the most offensive offal, such as decaying seal-skins, rancid seal, and whale blubber, &c. When on board my ship, they ate or drank greedily whatever was offered to them, salt-beef, salt-pork, preserved meat, pudding, pea-soup, tea, coffee, wine or brandy—nothing came amiss.

Of the Fuegians' typical shelter, which Europeans generally called "wigwams" and characterized as "the last degree of wretchedness," Stokes again was not content with second-hand descriptions but brought to them his own accurate eye.

> To their dwellings have been given, in various books of voyages, the names of huts, wigwams, &c; but, with reference to their structure, I think old Sir John Narborough's term for them will convey the best idea to an English reader; he calls them "arbours." They are formed of about a couple of dozen branches, pointed at the larger ends, and stuck into the ground round a circular or elliptical space, about ten feet by six; the upper ends are brought together, and secured by tyers of grass, over which is thrown a thatching of grass and seal-skins, a hole being left at the side as a door, and another at the top as a vent for smoke.

In other words, like the North American Indians (as Europeans, with the lingering cultural memory of the motivations of the first westbound explorers, still referred to aboriginal peoples everywhere west of the Atlantic), the Fuegians had evolved methods and techniques well-adapted to their environment and climate. But this was rarely appreciated by Europeans, who invariably interpreted what they saw as squalor and ungodly sloth.

To a degree greater than anyone else on first acquaintance, Stokes saw in them some of the sweeter traits of the universal human family.

> Their manner towards their children is affectionate and caressing. I often witnessed the tenderness with which they tried to quiet the alarms our presence at first occasioned, and the pleasure which they showed when we bestowed upon the little ones any trifling trinkets. . . . I took a fancy to a dog lying near one of the women . . . and offered a price for it. . . . She declined to part with it. . . . at last my offers became so considerable, that she called a little boy out of the thick jungle (into which he had fled at our approach), who was the owner of the dog. The goods were shown to him, and all his party urged him to sell it, but the little urchin would not consent.

They were all too human. And the history of their relationship with the technologically advanced white men who came from over the horizon followed the same ineluctable course it did everywhere else. Years of trading with sealers had given them a taste for, in the beginning, beads and mirrors, and later the more useful things: metal, cloth, tools, and weapons. In North America, the single most useful item introduced to the natives at their first contact with Europeans, the Spanish conquistadores, was the horse. That tool brought them speed, efficiency, and power; it altered their view of their world and what they could do in it. It was the transforming bone tossed into the air by Stanley Kubrick's ape that turns into a spaceship.

In the waterworld of Tierra del Fuego, what the Fuegians coveted most were the white men's boats.

To the elegant, aristocratic, accomplished Robert FitzRoy, looking down at the naked, greasy primitives in their canoes from the immeasurably loftier height of the *Beagle's* deck, the Fuegians seemed at first no more than curious local fauna. They bore no relation to him or his work. He was there to survey for the British Admiralty, to employ the formidable skills he had developed and make a name for himself in the world he knew. Yet it was a profoundly fateful encounter. His life and unique place in history, and the entire arc of scientific and religious thinking in the nineteenth century, were to turn on the meeting between Robert FitzRoy and these "savages."

It was the southern autumn when the survey ships reached Tierra del Fuego. The *Beagle*, and the schooner *Adelaide*, now commanded by Lieutenant Skyring, were sent to explore the western half of the Strait of Magellan.

Unlike Stokes, FitzRoy admired the landscape: "I cannot help here remarking that the scenery this day appeared to me magnificent," he wrote in his journal on April 14, near Port Famine. He so remarked on many days.

In May, FitzRoy anchored the *Beagle* and set out with some of his crew and a month's supplies in the ship's small cutter and a whaleboat, both of which could be easily rowed and sailed, and in these poked far into small bays and narrow channels where the larger ship could have sailed only with difficulty. This small mobile expedition traveled for more than a month, the men camping ashore at night. Following a small, twisting channel that led north from the strait through high snow- and ice-covered hills, they discovered and partially surveyed two vast sea-lakes, each about forty miles long and connected by a narrow channel, hidden away in the southern Andes. FitzRoy named these Otway Water, after his patron, and, with a nod to his loyal subordinate who had gracefully made way for him, Skyring Water.

He reveled in the rigors of the small-boat adventure. As the season advanced and the weather grew colder, he found that his navy cloak, which covered him at night, was stiff with frost in the mornings. "Yet I never slept more soundly nor was in better health." Late one afternoon, FitzRoy and his crew in the whaleboat were caught in a sudden gale and spent five hours of darkness rowing into the rising wind and waves, bailing frantically to keep the boat from swamping, until the gale died down just as quickly and they made shore. This incident, and the hardship shared and even enjoyed by FitzRoy and his men (with whom he also shared his cutthroat razor), made a strong bond between the new captain and his crew.

In July, with gales bringing snow and winter coming on, the *Beagle* and the *Adelaide* sailed west out of the strait, into the Pacific, and north to Chiloé Island to rendezvous with Captain King and the *Adventure*. On the way, scudding before a southeasterly gale, a moment's inattention by the helmsman allowed the *Beagle* to slew sideways off course and a breaking sea smashed aboard and swept away one of the whaleboats hanging in davits off the ship's stern quarter.

The three ships and their crews spent the southern winter anchored off a small settlement on Chiloé Island, working on the ships, readying them for the next season's work. Jonathan May, the *Beagle*'s carpenter, built a new whaleboat to replace the one that had been carried away. May was a shipwright of considerable skill who had served a long apprenticeship in boatbuilding, and the new boat was built of seasoned planks that had been carried aboard the *Beagle* from England expressly for such a need. FitzRoy was pleased with the new whaleboat. He did not know it was destined to act as the lynchpin of his fate.

In November, with the approach of spring, Captain King gave FitzRoy his orders for the coming season: alone with the *Beagle* (the *Adelaide* would be otherwise occupied), he was to explore and survey whatever he could of the ragged, broken southern shores of Tierra del Fuego, from the western to the eastern entrances of the Strait of Magellan. Neither captain knew the magnitude of these orders. Neither had any idea of the extent of this nebulous region's vast galaxy of razor-sharp rocks and islands scattered across more than 25,000 square miles, bordered by the strait to the north and Cape Horn to the south. FitzRoy had a few inadequate charts, and these were so full of small starlike crosses marking rocks that they appeared (as they still do on modern charts) more like maps of heavenly constellations. They were inaccurate, but they were good indicators of the nature of what lay ahead.

King told FitzRoy to accomplish what he could and rendezvous with the *Adventure* either in Port Famine by April 1 or in Rio de Janeiro by June 20.

The *Beagle* sailed south on November 19, 1829.

4

Small wonder the job had driven Pringle Stokes mad.

Sailing a bulky, square-rigged vessel through a place that invited shipwreck at any moment from uncharted rocks, hurricane-force williwaws, racing tides, obscuring blizzards (in any month of the year)—a place ship captains then as now considered the ultimate test of their seamanship and ability to manage a vessel, an ordeal to be got through and left behind—that was just the beginning.

The surveyor-captain, FitzRoy's inherited job, had to stop and deliberate in such a place. To find a tenable, secure anchorage, moor his ship, take readings—compass azimuths and angles shot from on deck with a sextant of landmarks and heights around the anchorage; then attempt a landing ashore through swell and surf in a small boat loaded with precious and delicate surveying equipment, scale a rocky hill or cliff with these instruments—a heavy theodolite, its tripod, sextant, and compass—take multiple readings; and do this many times in the immediate area to produce a single result of any useful accuracy.

From FitzRoy's journal, November 1829:

27th. A promising morning tempted me to try to obtain observations and a round of angles on or near Cape Pillar. I therefore left the ship with the master . . . [in the hands of the ship's sailing master, Mr. Murray, to tack back and forth off Cape Pillar] and went in a boat to the Cape. To land near it in much swell was not easy upon such steep and slippery rocks; at last we got inshore in a cove, and hauled the instruments up the rocks by lines, but could get no further, on account of precipices; I, therefore, gave up that attempt and went outside the Cape, to look for a better place; but every part seemed similar, and, as the weather was getting foggy, it was useless to persevere. . . .

28th and 29th. Gloomy days, with much wind and rain; and the gusts coming so violently over the mountains, that we were unable to do any work, out of the ship.

Whole weeks might pass like this: strenuous, dangerous efforts made for no result. The mission stalls and loses shape, goals are abandoned and redefined out of what is decreasingly possible. One's career—not to mention the lives and safety of the ship's complement—hangs in the diminishing balance.

Often, the problem was not being wrecked ashore but simply getting close enough to it.

Shortly after, the weather became so thick, that I could not see any part of the coast; and therefore stood offshore, under low sail, expecting a bad night. . . . The thick weather, and a heavy swell, induced me to stand farther out than I had at first intended. . . . After noon it was clearer, and we again stood inshore; but found that the current was setting us so fast to the southward . . . that we could not hold our own. . . . A good idea may be formed of the current which had taken us to the S.E., when I say that, even with a fresh and fair wind, it occupied us the whole of the [day] to regain the place we had left the previous evening. . . .

Dec 5th. To our mortification, we found ourselves a great way off shore; and Landfall Island, which was eight miles to leeward the last evening, was now in the wind's eye, at a distance of about six leagues [18 miles].

Sailors may know something of what FitzRoy felt that day at seeing such distance lost, and a course dead to windward over a foul current to make it back, in a ship that would point at best at an angle of 75 degrees toward its destination.

But unlike Stokes, FitzRoy was made for this. He was obdurate, determined, and resourceful. He quickly realized that sailing the *Beagle* alongshore, anchoring frequently, landing by ship's boat with instruments and then sailing on again was not going to work here. Too much time was being spent simply handling the weather and negotiating the ship's progress along the coast. It would be far better, he concluded, to anchor the *Beagle* in a safe harbor, leave her there, and move up and down sections of the coast by small boat, taking supplies enough to be independent of the ship for a week or more at a time—as he had done the previous season, north of the strait in Otway Water—then sail on to another base well along the coast. A minimum of time spent sailing the *Beagle*, more surveying in the small boats.

Through November and most of December, the *Beagle* cruised southeast along the southern shore of Desolation Island. This remote outlying bulwark edge of the world lay fully exposed to the storm seas and great swells of the Southern Ocean where they crashed at the end of their long run across the bottom of the Pacific. Yet even here, all along the coast, the Englishmen saw wigwams ashore and met with Fuegians.

Just before Christmas, they discovered a rare protected anchorage off Landfall Island, only forty miles from Cape Pillar, at the western entrance to the Strait of Magellan. Smoke indicated Fuegians on the shore. Soon after the *Beagle* anchored, a canoe full of men, women, and children, sixteen in all, approached the ship. The natives in it "were in every respect similar to those we had so frequently met before," wrote FitzRoy.

But they were not. They were members of the Yamana (or Yaghan, an English corruption) tribe, as they called themselves,

"Yammerschooner"—the endlessly importuning Fuegians near Button Island, Murray Narrows. (Narrative of HMS *Adventure* and *Beagle*, by Robert FitzRoy)

which meant simply "the people" in their own dialect. The Yamana may have resembled the Fuegians FitzRoy had seen farther north in the Strait of Magellan, but they were a tougher bunch, hardened and calloused from living along the wild stormwrack edge of the Southern Ocean. They were unimpressed by the beads and trinkets offered by the *Beagle*'s crew. They wanted knives, tools, metal, items the Englishmen wouldn't part with. "Yammerschooner" was the English transliteration of a word they heard constantly uttered by the Fuegians, here and throughout Tierra del Fuego, accompanied by the natives pointing to a coveted article. "Give me," they took it to mean.

But despite the nuisance of yammerschoonering Indians, FitzRoy liked his new anchorage: it offered good shelter yet was easy to sail into and out of, rare along this stretch of coast, and the bottom shoaled gradually and offered good holding for the *Beagle*'s anchors. Ashore, he obtained good sextant observations for latitude, and he named the place Latitude Bay.

On December 21, FitzRoy sent Mr. Murray, the ship's sailing master, off in a whaleboat with six other men and surveying instruments to the east side—the other side—of Landfall Island. The next day a gale began to blow from the northwest. On the twenty-second and twenty-third the storm intensified and rain drew a dense, obscuring curtain around the *Beagle*. Murray and his crew in the whaleboat had gone to the lee side of Landfall Island, which would be relatively sheltered from the storm, but they would clearly not be able to row or sail back against such weather to the ship.

Christmas Day and the twenty-sixth passed and the weather continued blowing and raining hard. FitzRoy grew anxious about the whaleboat crew. There was no possibility of sending a boat to look for them in such weather, and he knew they couldn't get back against it. He could only hope they were well and sheltered. They had set out with four days' provisions, not expecting to be gone that long, but they carried a good tent with them, and guns and ammunition enough to shoot plenty of the local "steamer" ducks, a flightless fishy-tasting waterfowl that propelled itself by paddling its short wings like a paddle steamer.

On the twenty-seventh, the wind finally moderated, and FitzRoy ordered all hands aboard the *Beagle* to keep a lookout for the whaleboat. At noon, two of the boat's crew were spotted waving to the ship from Landfall Island. A boat was lowered and they were soon brought back aboard. They had spent the previous afternoon and night walking across the island. The rest of the whaleboat's crew were still in a cove on the east side. Food finished, ammunition soaked, they had not been able to make a fire or shoot any game, and had eaten nothing for the past two days. On almost every day they had tried to row back around the island to the ship, but the weather had either forced them back ashore or threatened to blow them out to sea. Not knowing when the weather might let up, Murray had sent the two men across the island to try to get word to the ship. And as they came down to the shore of the *Beagle*'s anchorage that morning of the twenty-seventh, before they were seen by the ship, a group of Yamana Fuegians had attacked the two seamen, beaten them, and taken some of their clothing.

FitzRoy was astonished at such boldness by the natives and was determined to discourage it.

By this time, the weather was fine and before a rescue party reached them, the whaleboat appeared and Murray and the rest of the boat's crew made their way back to the ship.

FitzRoy and some of his officers went ashore to look for the offending Fuegians—and found them: an encampment of about twenty, of whom eight were men. They met the Englishmen armed with clubs, spears, and "swords," wrote FitzRoy, "which seemed to have been made out of iron hoops, or else were old cutlasses worn very thin. . . . They must have obtained these, and many trifles we noticed, from sealing vessels. By the visits of those vessels, I suppose, they have been taught to hide their furs and other skins, and have learned the effects of fire-arms."

FitzRoy does not say in his journal what happened at this confrontation. He slides abruptly away from the vivid and tantalizing picture of the natives waiting confidently for the Brits with clubs and cutlasses, into a bland, pedagogic description of their food: "the chief part of their subsistence on this island appeared to be penguins, seal, young birds." These Fuegians were not a pliant, submissive group. The situation had all the makings of a bloody skirmish, and the Englishmen wisely turned tail and walked away. But a corner had been turned.

A month later, on January 29, 1830, the *Beagle* anchored in a cove (today called Puerto Townshend) on London Island, which FitzRoy wanted to survey. Again he sent Murray and a crew off in a whaleboat to make instrument observations, this time of the area around a high craggy cliff about fifteen miles away, named Cape Desolation by Captain Cook fifty years earlier. Very soon after Murray and his men sailed off, the wind began to increase rapidly. By evening it was blowing what FitzRoy described in his journal (after almost a year in these storm-swept waters) as a "hard" gale. A note of worry about the men in the whaleboat again crept into his journal, but he took comfort from the fact that Murray knew what he was doing, and that "he could not have been in a finer boat."

It was the whaleboat newly made only months earlier by the ship's carpenter, Jonathan May, at Chiloé Island.

This gale continued to harden. The high peaks immediately above the anchorage made it worse. They funneled the wind into the furious katabatic blasts for which the Cape Horn region is so famous. Hurricane-force squalls tore down the slopes into the cove: "the williwaws were so violent, that our small cutter, lying astern of the ship, was fairly capsized. . . . the ship herself careened, as if under a press of sail, sending all loose things to leeward with a general crash."

The weather dismayed FitzRoy.

Considering that this month corresponds to August in our climate [the latitude of London Island, at 54 degrees, 40 minutes South, corresponds to Yorkshire in the northern hemisphere], it is natural to compare them, and to think how hay and corn would prosper in a Fuegian summer. As yet I have found no difference in Tierra Del Fuego between summer and winter, excepting in the former the days are longer, and the average temperature is perhaps ten degrees higher, but then there is also more wind and rain.

This was no England: Winter (July) temperatures averaged 31 degrees Fahrenheit; in midsummer's February they rose to the forties.

Five days passed.

The gale still continued, and prevented anything being done out of the ship. However safe a cove Mr Murray might have found, his time, I knew must be passing most irksomely, as he could not have moved about since the day he left us.

More than irksomely. On their first day Murray and his men had pulled into a cove in the lee of Cape Desolation, moored the whaleboat off the beach, and bivouacked ashore. At two o'clock the next morning, Murray sent one of his men to check on the boat; he reported back that it was riding happily at its mooring. At four o'clock another man got up to look at the boat and found it was gone.

At first they thought it must have blown away so they spread out alongshore hoping to find it drifting nearby. Then they discovered a Fuegian campsite of deserted wigwams and a still-burning fire. The boat had been stolen, they concluded, the thieves gone with it.

Murray set his men to building a makeshift boat to get word back to the *Beagle*. With considerable ingenuity, they fashioned something resembling an Irish coracle, a wicker-like intertwining of branches, covered with pieces of canvas cut from their tent, the inside packed with dense, clayey dirt. But they had to wait five days before the weather moderated enough to put to sea. The men had taken their clothes, the surveying instruments, and some of their provisions out of the boat the night before it was stolen, and they ate what food there was while they waited out the weather.

Early on February 4, two of the men set out with a ship's biscuit each for the fifteen-mile paddle to windward. Twenty hours later, at three in the morning of the fifth, exhausted, hungry, their wicker boat nearly sinking, they heard the barking of one of the dogs aboard the *Beagle* and found their way into the cove and to the ship. The men on the *Beagle* were amazed they had come so far in what appeared to them all to be nothing more than a large basket.

FitzRoy lost not a moment in going after the stolen boat. Into another five-oared whaleboat—also newly made by carpenter May at Chiloé—he piled two tents and two weeks' provisions for eleven men. While it was still dark he set off with six others to Cape Desolation. They reached Murray and his three companions just before midday. FitzRoy immediately asked to be shown where and how the lost boat had been moored—he still couldn't believe the natives had the nerve to take it; far more likely, he thought, it had broken away from its mooring in the gale and been blown out to sea. But when he saw the protected mooring place, and heard from his sailing master, in whom he had a nearly unwavering trust, how the boat had been secured, he was convinced of Murray's story.

The eleven men climbed into the whaleboat and set off on what would prove to be one of the most fateful pursuits in history.

5

FitzRoy has been characterized as going after the Fuegians who stole the whaleboat out of anger, pique, even imperial arrogance, to punish the natives for their theft, to teach them a lesson. He may have been angry, even furious, but the enormity and difficulty of his task, the length of the coastline assigned to him to survey, the constantly delaying weather, precluded the possibility of any side trip other than one facilitating the absolute need to continue his work.

A ship's boats comprised by far the most vital part of her gear. They enabled the crew to get ashore, to fuel and reprovision, and, if the ship foundered, offered the only hope of survival. For survey work, particularly as FitzRoy was now using the boats to explore islands and channels while the ship remained at anchor, they were as essential as a sextant or theodolite. The *Beagle*'s commission could not have been carried on without numbers of them.

While she was away from England between 1826 and 1830, the *Beagle* generally carried six boats, and she needed every one of them. The yawl, the largest at twenty-eight feet long, nearly a third of the ship's length and probably weighing three tons, had been smashed by a wave while being towed astern off the Patagonian coast the year before, deepening Pringle Stokes's depression and sense of failure. That left the ship's cutter, a 23-foot rowing and sailing gig; the jolly boat, 14 feet long; and usually three whale-boats, about 25 feet long, the same fast-rowing peapods carried aboard whaling ships. The jolly boat, a sort of general-purpose dinghy, hung over the stern in davits. The cutter and a whale-boat hung over each stern quarter. Two more whaleboats and the yawl, until it was lost, were carried on deck, taking up con-siderable room and restricting movement of the crew. While sev-eral whaleboats might be away from the ship for days at a time on survey work, all the other boats would have been in constant use going between ship and shore. "The people employed wood-ing ashore" was a daily recurring entry in the *Beagle*'s logbook whenever she was at anchor, referring to the shore party felling and chopping trees and gathering dried wood for the incessant demand of the ship's galley, forge, and other fires. Additional groups would be out every day—weather permitting—hunting and fishing to augment the basic rations of salted beef and ship's biscuit. The *Beagle* carried its own small, constantly busy fleet to serve its needs, and there was no room aboard for superfluous or extra craft. Every boat was vital. For survey work, none sur-passed the efficiency of the light whaleboats: "Our cutter required too many men, and was neither so handy, nor could she pull to windward so well as a whaleboat."

The *Beagle* had left England with enough seasoned planking for its carpenter to make several new boats, and in time he used up every foot of it, for in four years six boats were lost. By now, early 1830, there was not enough wood left in the ship's hold to build new boats. FitzRoy could not afford the loss of another. He needed it, so he set off with the greatest determination to get it back.

> The very first place we went to, a small island about two miles distant, convinced us still more decidedly of the fate of our lost boat, and gave us hope of retrieving her: for near a lately used wigwam, we found her mast, part of which had been cut off with an axe that was in the boat.

After finding the first signs of the stolen boat, they pulled and sailed (the whaleboats had short collapsible masts and sails to be used when the wind allowed) northeast, into a large bay dotted with many islands. Toward dusk, they drew level with a canoe being paddled by two Fuegians, a man and a woman. They indicated to the Englishmen, by signs, that they'd seen several boats heading into the northern part of the bay. "This raised our hopes, and we pushed on," FitzRoy wrote.

He may well have been angered by the theft of his boat, but he was clearly not disposed to think too badly of Fuegians in general.

> The woman . . . was the best looking I have seen among the Fuegians, and really well-featured: her voice was pleasing, and her manner neither so suspicious nor timid as that of the rest. Though young she was uncommonly fat, and did justice to a diet of limpets and muscles. Both she and her husband were perfectly naked.

After two days of fruitless searching, they came across a native family in two canoes near the head of the bay, thirty miles east-northeast of Cape Desolation. Something in these Fuegians' attitude prompted the Englishmen to search their canoes, which they had not done to the paddling naked couple. In one of the canoes they found the lost whaleboat's leadline.

> We immediately took the man who had it into our boat, making him comprehend that he must show us where the people were, from whom he got it. He understood our meaning well enough.

This was all FitzRoy wrote about his first taking of a hostage. For him it was an act that required no justification. It was a quick, practical decision, a tactic born of the necessity of the situation, but it was a signal moment of change in FitzRoy's relationship with the Fuegians.

It was probably accomplished by implicit rather than actual

force. The native canoe was being held alongside the much larger whaleboat while the sailors searched it. At a word from FitzRoy, two or three uniformed Royal Marines would have risen and, with a leg in each boat, "helped" the Fuegian who "had" the leadline into the whaleboat. With signs, FitzRoy would have indicated what he wanted, and the Fuegian "understood our meaning well enough." Then the Englishmen, with their captive, pulled away from the canoes.

The Fuegian led them to a cove containing a camp with wigwams and two more canoes on the beach, a third being built. At the sight of the Englishmen, the Fuegians ran into the nearby bushes with as many of their belongings as they could carry, then returned, empty-handed and naked, and huddled together on the beach.

FitzRoy's men found more of the missing boat's gear—a shredded piece of sail, an oar that had been broken in two (the shaft hacked into a seal club, the blade used as a paddle), and the boat's axe and toolbag. FitzRoy was convinced he had found the group responsible for the theft. Apart from an old man and a boy aged about seventeen, there were only women in the camp. Their men, he reasoned, were away in the whaleboat on a seal hunt.

He took a second captive, who may or may not have joined the Englishmen as cheerfully as FitzRoy described.

> The women understood what we wanted, and made eager signs to explain to us where our boat was gone. I did not like to injure them and only took away our own gear, and the young man, who came very readily, to show us where our boat was, and, with the man who had brought us to the place, squatted down in the boat apparently much pleased with some clothes and red caps, which were given to them.

With their two hostages providing directions, the Englishmen rowed away. They pulled hard against a rising wind through the long twilight, heading deep into Courtenay Sound, a star-shaped bay of many fjordlike arms surrounded by high snowy hills. Four hours later, too dark to go on, they beached the boat for the night and made camp.

The two Fuegians seemed, to FitzRoy's eyes, quite at ease, so he decided not to "secure our guides as prisoners" for the night but let them sleep near the fire while the man on watch kept an eye on them. But in the predawn dark, they slipped away into the bushes, naturally taking with them the two tarpaulin coats they had been given to sleep under.

With daylight, the Englishmen rowed back along shore looking for their runaways. They returned to the "boat stealers' family" camp where they had taken the second hostage the day before. Again the Fuegians took to the woods as they approached. The Englishmen landed and destroyed the natives' canoes—to prevent news of their search for the stolen whaleboat traveling beyond the immediate area, FitzRoy wrote, but this act reeks of vengeful frustration.

For the next few days they rowed and sailed as best they could around the protected arms of Courtenay Sound while a strong gale blew from the south. They found nothing. FitzRoy decided to return once more to the "boat stealers' family" camp, but this time to take them by surprise and capture as many hostages as possible for the return of the stolen whaleboat.

The Fuegians had quite sensibly gone. But scouting from a hill the Englishmen spotted them: they'd moved their camp to another cove. The attack was planned for the following day.

Not knowing if the family's absent men had returned, FitzRoy armed each of his ten sailors and marines with a pistol or musket, a cutlass, and a length of rope to secure a prisoner. When morning came, they crept through the bushes toward the cove. They had nearly surrounded the camp when the Fuegians' dogs smelled them and began barking. The Englishmen rushed the camp.

At first the Indians began to run away, but hearing us shout on both sides, some tried to hide themselves by squatting under the banks of a stream. . . . The foremost of our party, Elsmore . . . in jumping across this stream, slipped, and fell in just where two men and a woman were concealed: they instantly attacked him, trying to hold him down and beat out his brains with stones; and before any one could assist him, he had received several severe blows, and one eye was almost destroyed by a dangerous stroke near the temple. Mr Murray, seeing the man's danger, fired at one of the Fuegians, who staggered back and let Elsmore escape; but immediately recovering himself, picked up stones from the bed of the stream . . . and threw them from each hand with astonishing force and precision. His first stone struck the master with much force, broke a powder-horn hung round his neck, and nearly knocked him backwards, and two others were thrown so truly at the heads of those nearest him, that they barely saved themselves by dropping down. All this passed in a few seconds, so quick was he with each hand; but, poor fellow, it was his last struggle; unfortunately he was mortally wounded, and, throwing one more stone, he fell against the bank and expired.

After some struggling, and a few hard blows, those who tried to secrete themselves were taken, but several who ran away along the beach escaped. So strong and stout were the females, that I, for one, had no idea that it was a woman whose arms I and my coxswain endeavoured to pinion, until I heard some one say so. The oldest woman of the tribe was so powerful that two of the strongest men of our party could scarcely pull her out from under the bank of the stream.

The Englishmen had bagged eleven prisoners—two men, three women, six children—among them the young man taken from the camp several days earlier. And when the dead, defiant, furiously stone-throwing Fuegian was examined, they recognized their first hostage: the man taken from the canoe because he had appeared to be in possession of the missing boat's leadline; who had seemed so happy with clothes and a red hat.

FitzRoy may have misread the Fuegians' docility, but he felt genuine remorse at having killed one. "That a life should have been lost in the struggle, I lament deeply; but if the Fuegian had not been shot at that moment, his next blow might have killed Elsmore, who was almost under water."

It was the first record of such a death at British hands, as FitzRoy surely knew, and this undoubtedly distressed him. He was there for survey work, and killing the locals was an unsanctioned departure from his job description, however he might justify it. It would not commend him to his superiors. But there was more to his regret than that, as his later actions would prove.

The prisoners appeared anxious to tell the Englishmen where the missing boat was, pointing now in another direction, to the southeast, not to Courtenay Sound. But with twenty-two people in a 25-foot whaleboat, FitzRoy was not going on another long chase. They headed instead for the *Beagle*, reaching it two days later, on February 15. The hostages were fed and clothed, and the *Beagle* weighed anchor and sailed southeast to Cape Castlereagh, in the direction the hostages had last indicated, and also where his survey might continue.

On February 17, FitzRoy and Murray set out to search again, in two boats, with a week's provisions and Fuegian hostage-guides in each, including two stout women, mothers of children left aboard the *Beagle*. "As far as we could make out, they appeared to understand perfectly that their safety and future freedom," and the safety of their children aboard the ship, "depended upon their showing us where to find the whaleboat."

Tantalizingly, in the first cove he came to, only two miles from where the *Beagle* was now anchored, FitzRoy found another piece of the missing boat's leadline in a "lately deserted" wigwam. They found more signs of a large party of Fuegians among several islands nearby, and he became hopeful that he would soon find his stolen boat. They camped ashore, and again FitzRoy decided not to tie up his prisoners for the night, reasoning that the children back aboard the *Beagle* would bind the women more securely than any rope.

> I kept watch myself during the first part of the night, as the men were tired by pulling all day, and incautiously allowed the Fuegians to lie between the fire and the bushes, having covered them up so snugly, with old blankets and my own poncho, that their bodies were entirely hidden. About midnight, while standing on the opposite side of the fire, looking at the boats, with my back to the Fuegians, I heard a rustling noise, and turned round; but seeing the heap of blankets unmoved, satisfied me . . . another rustle, and my dog jumped up and barking, told me that the natives had escaped. Still the blankets looked the same, for they were artfully propped up by bushes.

For another week the two boats searched the Stewart and Gilbert Islands, a labyrinth of coves and channels where they now believed the stolen whaleboat might be. They saw fires, found deserted camps, they even saw Fuegians running off at their approach, but no boat. They finally returned to the *Beagle* on February 23, to learn that all the ship's hostages, except for three children, had escaped.

> Thus, after much trouble and anxiety, much valuable time lost . . .
> I found myself with three young children to take care of, and no prospect whatever of recovering the boat.

But the search for the boat, the continuing attempt to second-guess its thieves, the constant confoundment of all his expectations of the Fuegians' behavior, had sown in FitzRoy a seed that would grow to a major preoccupation. For weeks, and then months, references to his surveying work—his sole purpose for being there—faded from his journal. Observations on harbors, weather, seafaring, diminished until they were almost entirely eclipsed by his mounting interest in the Fuegians:

This cruise had . . . given me more insight into the real character of the Fuegians, than I had then acquired by other means. . . . I became convinced that so long as we were ignorant of the Fuegian language, and the natives were equally ignorant of ours, we should never know much about them, or the interior of their country; nor would there be the slightest chance of their being raised one step above the low place which they then held in our estimation.

In this practical observation, with its telling grammatical tense, lie all the imperial ambitions of England at the time when FitzRoy wrote it—not in February 1830 when he was looking for his lost whaleboat, although it is set down as a journal entry for that time, but seven or eight years later, when he was preparing his journals for publication. After Victoria had acceded to the throne in 1837, her love of India, as a prize, and of the concept of empire, was fueling British imperial expansion across the globe. When FitzRoy wrote, in 1837 or 1838, of uplifting the heathen savage while at the same time gaining knowledge of the interior of his country, he was tapping into the major preoccupation of the age to explain and justify the deepening of his own obsessive fascination for the Fuegians, and the turn this was about to take in 1830.

Late in February, the *Beagle* sailed on down the ragged southeastern coast of Tierra del Fuego and anchored in Christmas Sound, "in the very spot where the *Adventure* lay when Cook was here," FitzRoy wrote in his journal.

He was mistaken, confusing ships. The *Adventure*, under the command of Tobias Furneaux, had left England in company with Captain James Cook's *Resolution* in July 1772, on Cook's second great circumnavigation of exploration. The two ships cruised partly in company as far as New Zealand, but they lost touch with each other in 1773. In 1774, Cook reached Tierra del Fuego for the second time, after an icy, high-latitude crossing of the South Pacific from New Zealand, searching for the mythical Terra Incognita, which he concluded did not exist. (He missed seeing Antarctica by only a few hundred miles.) In mid-December 1774, *Resolution* closed with Tierra del Fuego near the western entrance to the Strait of Magellan, passing a headland that Cook named Cape Gloucester. For two days, *Resolution* scudded on southeast before a westerly gale (along the same track followed by FitzRoy in the *Beagle* while surveying and looking for the lost whaleboat) until it neared a black 800-foot-high rocky promontory rising from the sea, which Cook named York Minster after the great cathedral in his home county of Yorkshire. Here a southeasterly breeze stopped him, and he turned his ship into a channel and found shelter. *Resolution* was anchored over Christmas while Cook surveyed and charted the surrounding coastline, naming the area Christmas Sound, and so it is still called today.

Into this spot, fifty-five years and two months later, came FitzRoy in the *Beagle*. "His [Cook's] sketch of the sound, and description of York Minster, are very good, and quite enough to guide a ship to the anchoring place."

Just east of York Minster was a more protected anchorage, mentioned by Cook, but not examined or named by him, and into this sheltered cove FitzRoy worked the *Beagle* on March 1, naming it March Harbour. It seemed a good place to leave the ship for several weeks while he and Murray again set out in two different boats, no longer on a wild goose chase, but to continue their mission with surveying instruments. And here FitzRoy set carpenter May to building another whaleboat. Since there was no longer enough planking for this in the ship's stores, May cut up a spare spar, another ship's former topmast that was being saved as a replacement for the *Beagle's* main topmast. "With reluctance this fine spar . . . was condemned to the teeth of the saw; but I felt certain that the boat Mr May would produce from it, would be valuable in any part of the world, and that for our voyage it was indispensable."

FitzRoy dispatched Murray in the cutter to survey the coast, channels, and islands to the west—back toward the area they had searched for their stolen whaleboat. He sent with him two of the three children remaining aboard the *Beagle*, to be left with any Fuegians he found.

> The third, who was about eight years old, was still with us: she seemed to be so happy and healthy, that I determined to detain her as a hostage for the stolen boat, and to try to teach her English.

The *Beagle*'s crew, in a nod to the wicker-like craft built by Murray and his men that had first brought them news of the stolen whaleboat, had taken to calling this child Fuegia Basket.

6

FitzRoy's instructions from the British Admiralty contained no provisions about capturing or killing foreign nationals. He sailed a warship across a lawless world and what he did was up to him. His code of behavior was that of an English gentleman, which carried with it the assumption of moral, intellectual, and religious superiority. This gave him, he felt, the unquestioned right to attempt to retrieve his stolen property as he saw fit, to stop, question, capture, and even kill natives in the course of his inquiries, if this seemed necessary. He tried not to kill, he behaved as decently as he thought fit, but kidnapping people didn't faze him. He took to it without hesitation.

The acquisition of Fuegia Basket marked a turning point. It was by then surely clear to him that hostage-taking was unlikely to produce the ransom of his missing whaleboat. Fuegian mothers, pretending to lead the crew toward their missing boat, had run away, effectively abandoning their children held aboard the *Beagle*, rather than complying with the Englishmen's demands. No entreaty had come for Fuegia Basket, child of the "boat stealers' family." She was unclaimed property.

Clothed in seaman's garb, the little girl had the run of the ship. Eight years old (young enough to lie below the sexual radar of most of the *Beagle*'s crew), small, easily amused, no doubt amusing, she had become, according to FitzRoy, "a pet on the lower deck." From his earliest descriptions of her, FitzRoy was keenly aware of her as a personality; he saw the child rather than her use as a bargaining commodity. He was charmed by her. He wanted to keep her.

As carpenter May worked ashore in March Harbour on the new whaleboat, FitzRoy and Murray tried to turn their attention once more to surveying, but Fuegians again got in the way. A group of them in a canoe approached the ship on March 3 wanting to come aboard. Impatient with the nuisance they represented, wary of their pilfering with May's carpentry shop set up on shore, FitzRoy sent Mr. Wilson, the mate, in one of the boats to chase them away, to fire pistol shots over their heads.

But almost immediately, his curiosity about the Fuegians, by now deepened nearly to obsession, made him change his mind. He set out himself in another boat. In his published journals, FitzRoy would later write: "Reflecting . . . that by getting one of these natives on board, there would be a chance of his learning enough English to be an interpreter, and that by this means we might recover our lost boat . . . I went after them, and hauling their canoe alongside of my boat, told a young man to come into it; he did so, quite unconcernedly, and sat down, apparently contented and at his ease." The rest of the Fuegians "paddled out of the harbour as fast as they could."

Back aboard the *Beagle*, the young man was christened York Minster, after the dominating topographical feature of the neighborhood. He was cleaned and fed, and introduced to Fuegia Basket. They talked and York Minster, sullen at first, grew "much more cheerful."

Five days later, while on a hill taking angles for his survey of March Harbour, FitzRoy saw smoke coming from a cove near the harbor entrance. Unable to resist, he ran down to the shore and had two men row him to the cove to see if this group possessed anything that might have come from the stolen whaleboat. But as the Englishmen approached, these natives became "very bold and threatening," so FitzRoy returned to the *Beagle*, filled two boats with armed men, and set off after the Fuegians, who were now paddling away fast across the harbor. The Englishmen chased them to shore where a fight ensued, an exchange of rocks and gunfire, during which no one was injured except for a seaman hit by a rock. The natives escaped into the bush. FitzRoy's men found part of the lost boat's gear in the beached canoes, and he concluded that among this group must be the whaleboat's thieves. (Items from the stolen boat—oars, line, beer bottles—appeared to have been so widely disseminated throughout Tierra del Fuego that FitzRoy was able to see the thieves everywhere.) He destroyed their canoes.

The next morning he set out again with an armed party in the direction of smoke seen above nearby islands, hoping still to find his boat. Again he saw Fuegians paddling away in canoes and intercepted them. As the Englishmen's boat reached the first canoe, its occupants jumped overboard. The crew grabbed one of them, a young man, who was hauled into FitzRoy's boat after a fierce fifteen-minute struggle in the water. The Englishmen returned to the *Beagle* with this new captive, whom FitzRoy optimistically christened Boat Memory. Despite being frightened, the new Fuegian aboard "ate enormously, and soon fell fast asleep."

FitzRoy now had three Fuegian captives aboard the *Beagle*. He was as happy with them as a big game hunter with a good bag of trophies.

"Boat" was the best featured Fuegian I had seen, and being young and well made, was a very favourable specimen of the race; "York" was one of the stoutest men I had observed among them; but little Fuegia was almost as broad as she was high: she seemed to be so merry and happy, that I do not think she would willingly have quitted us. Three natives of Tierra del Fuego, better suited for the purpose of instruction, and for giving, as well as receiving information, could not, I think, have been found.

Some design for them, not fully formed, was taking shape in FitzRoy's mind. The specimens were clothed in regulation seaman's dress, and instructed in English.

With the new whaleboat completed by carpenter May, the *Beagle* sailed from March Harbour on the last day of that month. The ship trended southeast along the ragged shore of Tierra del Fuego. FitzRoy no longer chased native fires in search of his stolen boat and concentrated on his surveying work. His Fuegian specimens were, according to him, "becoming very cheerful, and apparently contented."

In Orange Bay, on the Hardy Peninsula west of Cape Horn, a group of Fuegians approached the ship to barter. FitzRoy was surprised at the reaction of his captives to these visitors. They spoke a different dialect than Boat Memory and York Minster, who nevertheless recognized the newcomers and yelled at them, calling them, FitzRoy believed, "Yapoo." They showed FitzRoy scars from wounds they'd received fighting the "Yapoos," a distinct and different tribe, he concluded. FitzRoy also referred to them as "Yahoos"—he had undoubtedly read *Gulliver's Travels* (1726), whose protagonist refers to the brutish and imaginary race of that name as "those filthy Yahoos."

Late in April, the *Beagle* anchored near Horn Island, the southernmost point of South America, the false cape, the infamous Ultima Thule to all seamen known as Cape Horn. FitzRoy and a party rowed ashore, climbed the island's height, and took the usual observations. Then they put aside their instruments and erected an 8-foot high pile of stones over a memorial to dead seamen, broke out the Union Jack, and toasted the health of King William IV. Like tourists everywhere, like the plundering Oxford scholars in Greece or the astronauts who visited the moon, when they rowed away from the island they took with them fragments of Cape Horn.

Early in May the *Beagle* anchored in a bay on the east coast of Lennox Island, north of Cape Horn. Three boats headed off to survey the area around wide Nassau Bay. FitzRoy went in one of them with a group of seamen, heading west across Nassau Bay and then north toward a narrow channel discovered by Murray on a boat trip a few weeks earlier. Murray Narrows, as FitzRoy named it, led into what appeared to be a wide straight channel that ran east and west through the heart of Tierra del Fuego, which FitzRoy called Beagle Channel.

FitzRoy met Fuegians in canoes and ashore, the same Yapoos encountered earlier. Unlike the aggressive, boat-thieving Yamana Fuegians farther west, these natives were mainly interested in barter. They had clearly had contact with sealing vessels, whose voracious appetite for every kind of skin made the Yapoos now attempt to hide their guanaco hides at the sight of the Englishmen. They offered instead fish, which they traded for beads and buttons. With one group, FitzRoy traded a knife for a "very fine dog."

On May 11, near the entrance to Murray Narrows, FitzRoy's boat was intercepted by three canoes eager for trade.

> We gave them a few beads and buttons, for some fish; and, without any previous intention, I told one of the boys in a canoe to come into our boat, and gave the man who was with him a large shining mother-of-pearl button. The boy got into my boat directly, and sat down. Seeing him and his friends quite contented, I pulled onwards, and, a light breeze springing up, made sail. Thinking that this accidental occurrence might prove useful to the natives, as well as to ourselves, I determined to take advantage of it. . . . "Jemmy Button," as the boat's crew called him, on account of his price, seemed to be pleased at his change.

With orders to be in Rio de Janeiro by June 20, FitzRoy turned the *Beagle* eastward to survey what remained of the coast of Tierra del Fuego before turning north to Rio and, beyond that, England.

Four Fuegian captives still remained aboard. FitzRoy now had a plan for them.

I had . . . made up my mind to carry the Fuegians . . . to England; trusting that the ultimate benefits arising from their acquaintance with our habits and language, would make up for the temporary separation from their own country. But this decision was not contemplated when I first took them on board; I then only thought of detaining them while we were on their coasts; yet afterwards finding that they were happy and in good health, I began to think of the various advantages which might result to them and their countrymen, as well as to us, by taking them to England, educating them there as far as might be practicable, and then bringing them back to Tierra del Fuego. . . . In adopting the latter course I incurred a deep responsibility, but was fully aware of what I was undertaking.

According to FitzRoy, the Fuegians "understood clearly when we left the coast that they would return to their country at a future time, with iron, tools, clothes, and knowledge which they might spread among their countrymen." The four natives could not possibly have comprehended such a scheme from a rudimentary exchange that might have communicated, at most, "You come with us, get tools, knives, we bring you back." But, according to FitzRoy, the only chronicler-witness to what was said and understood, the Fuegians appeared content and interested in self-improvement and made no use of several opportunities to escape.

They helped the crew whenever required; were extremely tractable and good-humoured, even taking pains to walk properly, and get over the crouching posture of their countrymen. When we were at anchor in Good Success Bay, they went ashore with me more than once, and occasionally took an oar in the boat, without appearing to harbour a thought of escape.

On June 7, the *Beagle* passed through the Strait of Le Maire, out of the waters of Tierra del Fuego, and sailed north into the warming Atlantic.

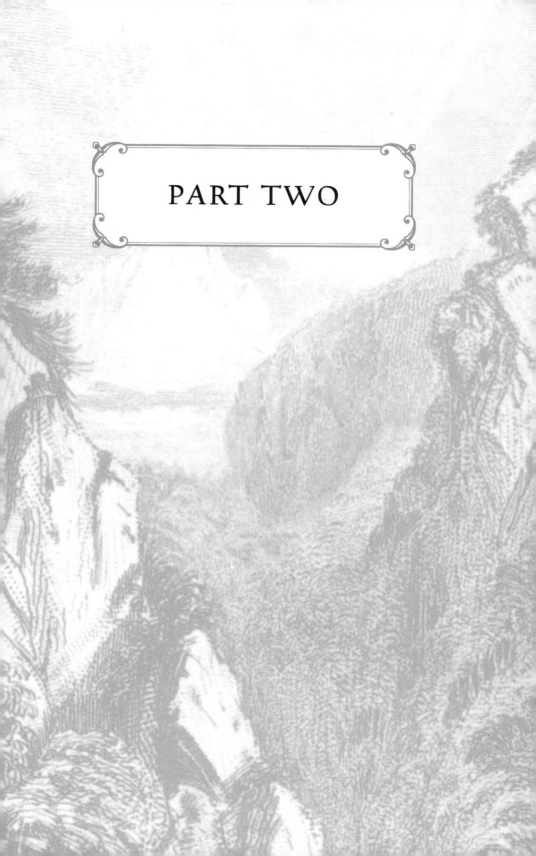

PART TWO

7

The term "collecting" had a particular weight attached to it
in nineteenth-century Britain. Explorers in Africa, Asia,
and South America felt duty-bound to collect bugs, birds, spiders, flowers, and native nose rings to ship home. Every far-flung
outpost of the empire had its local amateur enthusiast: the
Indian Raj hill station doctor who sent home a few butterflies,
the missionary in Africa who became fascinated with dung beetles, anyone of a scientific bent who gathered examples of the
local flora and fauna. He would look for them himself, and his
fellow expatriates and the local natives would bring him anything thought to be worthy of his interest.

These collectors would examine their specimens, categorize, preserve, then package the finer examples with care and send them
"home" to England. Sometimes they were sent to a friend who might
be storing, housing, or displaying the mounting collection, but just as
often to the British Museum or some other interested repository.
Country houses, museums, universities, and gentlemen's clubs filled
with specimens from around the globe. Taxidermy became a frenzied
profession. England, in that age of expanding exploration and colonial possession, became a vast storehouse of every kind of transportable evidence of the warp and weft of Man and Nature.

Notable collectors were Lord Elgin, who in 1806 looted a

shipful of 2,500-year-old marble statues from the Parthenon in Athens and sent them back to England, and naturalist Alfred Russel Wallace, whose eight-year-long wanderings through the Malay Archipelago eventually resulted in 125,660 specimens of plants, insects, and animals shipped home.

FitzRoy's savages were a natural part of this ethos. From the moment of first contact, Eskimos, Pygmies, Polynesians, Africans, and "Ioway" American Indians had also been "collected" by European explorers. Captain Cook returned to England at the end of his second voyage with a Tahitian, Omai, who partied in London for two years before returning home with Cook on his last voyage. Fifty-eight thousand people went to see a family of Laplanders with their live reindeer exhibited in London in 1822. Two hundred years earlier, Captain George Weymouth kidnapped five Indians from an island near what is now Port Clyde, Maine, and brought them back to England. They were treated well and when eventually repatriated, they had nothing but good to say of the English. One of them, Tasquantum, or Squanto, taught some English to a friend of his, Samoset, who happened to be in the neighborhood of present-day Plymouth, Massachusetts, when the *Mayflower* dropped anchor in December 1620. When its pilgrim passengers went ashore, they were met by Samoset who flabbergasted them by saying "Welcome" and asking if they had any beer.

Such kidnap victims were "specimens," as FitzRoy naturally described his Fuegians, like breadfruit or the Argentine opossum, to be collected and studied for the benefit of science—and for their own sake: what Kipling termed the white man's burden, the patronizing presumption that it was morally incumbent upon civilized Englishmen to extend (more often than not with overwhelming force) a hand to uplift their colored inferiors. Poked and prodded in all cases, the luckier specimens were treated with genuine compassion and occasionally sent back to their homes with chests of smart clothes and new belongings and amazing stories to tell their descendants. Others found themselves in smoky, overpopulated, industrial cities, exhibited as curiosities at fairgrounds and in theaters, examples of the freakish lower echelons of creation, destined to die of neglect, despair, and loneliness.

These living trophies satisfied the missionary zeal of the age that was the higher-minded rationale behind the scramble for colonial possession. Men like FitzRoy, and David Livingstone in Africa, believed they were bringing improvement and light to disenfranchised peoples while paving the way for the British to take over their lands and material wealth.

Whatever his motives, FitzRoy's collection of natives from the territory of his survey was not part of his job description. It was an action far beyond the strict orders plainly outlined by the Admiralty, whose responsibility for the Fuegians he had now incurred, to dubious purpose.

As the *Beagle* approached England in September 1830, FitzRoy wrote a letter to his superior officer, Captain Phillip Parker King, of the *Adventure*, telling him what he had done:

> *Beagle,* at sea, 12 September 1830
> Sir,
>   I have the honour of reporting to you that there are now on board of His Majesty's Sloop, under my command, four natives of Tierra del Fuego.
>   Their names and estimated ages are:
>   York Minster. 26
>   Boat Memory. 20
>   James Button. 14
>   Fuegia Basket (a girl). 9
>   I have maintained them entirely at my own expense, and hold myself responsible for their comfort while away from, and for their safe return to their own country: and I have now to request that, as senior officer of the expedition, you will consider the possibility of some public advantage being derived from this circumstance; and of the propriety of offering them, with that view, to His Majesty's Government.

FitzRoy then gave a brief account of the *Beagle*'s stolen whaleboat and his attempt to secure hostages and interpreters for its return, and his eventual decision to keep the captives aboard.

I thought that many good effects might be the consequence of their living a short time in England. They have lived, and have been clothed like the seamen, and are now, and have been always, in excellent health and very happy. They understand why they were taken, and look forward with pleasure to seeing our country, as well as returning to their own.

Should not His Majesty's government direct otherwise, I shall procure for these people a suitable education, and, after two or three years, shall send or take them back to their country, with as large a stock as I can collect of those articles most useful to them, and most likely to improve the condition of their countrymen, who are now scarcely superior to the brute creation.
Robt. FitzRoy

The passage from Tierra del Fuego was a long one, four and a half months, with stops in Montevideo and Rio de Janeiro. The Fuegians grew slowly more communicative with the crew, as they picked up a basic and sailorly English, and FitzRoy attempted to make translations of some of their words, a process complicated by the fact that Jemmy Button spoke a different dialect from the other three captives.

The Fuegians made rare tourists aboard their British Navy cruise ship. Montevideo, their first city, would have been a fantastic sight to them. The harbor was filled with ships, a sprawl of buildings and streets tumbled down to the water, large buildings and warehouses lined the docks, and music of all kinds poured out of bars and dance halls and floated across the water to the anchored ship. Here they all went ashore. FitzRoy took them to the local hospital to be vaccinated against smallpox; Fuegia Basket stayed with an English family for a few days, and the three men accompanied the captain (and probably several of the *Beagle*'s marines) on some of his business through the city.

The Fuegians seemed far less astonished and amazed than FitzRoy expected. Animals and boats—things they were familiar with—drew animated responses from them: "A large ox, with unusually long horns, excited their wonder remarkably." But with much else, the larger, denser jumble of civilization around them, the shock of the new, seemed to overload their senses and dull them into a stolid impassiveness.

> When anything excited their attention particularly they would appear at the time almost stupid and unobservant; but that they were not so in reality was shown by their eager chattering to one another at the first subsequent opportunity, and by the sensible remarks made by them a long time afterwards, when we fancied they had altogether forgotten unimportant occurrences which took place during the first few months of their sojourn among us.

In Montevideo, far from home, sweltering in a new, subtropical climate, and perhaps seeing the English captain for the first time as a friend and buffer between them and an unimaginably strange world, the Fuegians began to open up to FitzRoy and talk to him about their home and customs. "It was here that I first learned from them that they made a practice of eating their enemies taken in war. The women, they explained to me, eat the arms; and the men the legs; the trunk and head were always thrown into the sea."

As the known world opened up in the eighteenth and nineteenth centuries, cannibalism was the salacious bogey attached to every dealing with "savages." Fear of it had sent Captain Bligh and his fellow castaways from the *Bounty* more than 4000 miles across the Pacific and the Strait of Timor in an open boat, rather than putting ashore at any of the many islands and the Australian mainland they passed within sight of. Cannibalism was known to have been practiced in Africa, Polynesia, and Australasia; any dark-skinned race was presumed to be capable of it. Whether real or imaginary, it underscored the moral imperative that God-fearing Englishmen felt to improve the condition of native cultures everywhere.

According to Lucas Bridges, the son of a missionary who grew up in nineteenth-century Tierra del Fuego, there was never any cannibalism among the Fuegians. FitzRoy's misinformation, Bridges speculated in his book, *Uttermost Part of the Earth*, was probably the result of his own fearful and probing questions about cannibalism, which his Fuegian captives sensed and responded to with black humor: "Do your people ever eat each other?" "Oh, yes, the men eat the legs, the women the arms."

But with such expected disclosures from his protégés, FitzRoy was more certain than ever of the rightness of his mission to clothe them, carry them home, and steep them in Christian values.

After a brief stop in Rio de Janeiro, the *Beagle* sailed on August 6 and hove in sight of England nine weeks later. Montevideo, for all its newness to the Fuegians, was a ramshackle town at the edge of a wilderness; England was the center of the scientific world, throbbing, clamoring, and smoking in the full bore of industrial development. In Falmouth harbor, at the clean edge of the open sea, the Fuegians were terrified by their first sight of it.

> I think that no one who remembers standing for the first time near a railway, and witnessing the rapid approach of a steam-engine, with its attached train of carriages, as it dashed along, smoking and snorting, will be surprised at the effect which a large steam ship passing at full speed near the *Beagle*, in a dark night, must have had on these ignorant, though rather intelligent barbarians.

After dropping mail to be delivered to the Admiralty, the *Beagle* sailed on to Devonport, the naval dockyard at Plymouth.

Here FitzRoy took the Fuegians ashore at night to "comfortable, airy lodgings." The next day he brought a doctor to vaccinate them for a second time against smallpox.

A virus that spread like the common cold, smallpox had been more lethal to humans through the ages than all our wars. A victim expelled droplets containing the virus from the nose and mouth; anyone inhaling the droplets or carrying them by the hand to the mouth became infected. The symptoms were unmistakable: aches, a high fever, followed by a rash resembling thousands of small pimples on the face and spreading to other parts of the body. The pimples became larger and filled with pus. The disease killed about 20 percent of its victims. In those who survived, scabs formed over the pimples, leaving permanent scars. The eyes were often infected and many were left blind. Sweeping across Asia, Africa, and Europe, as influenza still does, smallpox was once so common that almost everyone got it at some time. Europeans unfailingly carried it with them in their explorations around the world, killing millions of natives who had no immunity to the disease.

Until the late eighteenth century, the only protection was variolation: inoculation of a healthy person with the pus of a smallpox victim. This could result in a mild case of the disease and subsequent immunity. In 1796, a British physician, Edward Jenner, went further and made a real vaccine from cowpox, a mild form of the disease suffered by milkmaids, who were then said to be immune to smallpox. As with John Harrison, the eighteenth-century clockmaker whose work was persistently ignored or rebuffed (but without which FitzRoy could not have determined his longitude with such accuracy), Jenner's claim and studies were sneered at by his own medical community and rejected by the Royal Society (Britain's premier association of scientists). But his vaccination was quickly adopted elsewhere. President Thomas Jefferson tried it out on family members; Napoleon vaccinated his troops; physicians in Europe and Russia began to use it. By 1830, when FitzRoy brought the Fuegians to England, vaccination was widespread, even, finally, in England (where Jenner had belatedly been rewarded with £30,000 by Parliament), but the vaccines were of varying quality. Plymouth, a seaport with a large and constantly revolving population of seamen and visitors, was a rich breeding ground for disease. FitzRoy was unsure of the quality of the vaccine at Montevideo, so he started his Fuegians on a new course of vaccinations as soon as possible.

Two days later he brought them to stay at a farm a few miles from Plymouth, where he hoped they would have more room and fresh air and less risk of disease, and where they could quietly remain while he figured out what to do with them.

This wasn't easy. John Barrow, the second secretary at the Admiralty, had responded favorably to his initiative: "Their Lordships will not interfere with Commander FitzRoy's personal superintendence of, or benevolent intentions towards these four people, but they will afford him any facilities towards maintaining and educating them in England, and will give them a passage home again." This meant little beyond accepting what FitzRoy had done and giving him permission to spend some of his time arranging the welfare of his charges. The Admiralty's only true "facilities" were the berths offered on its ships heading, at some future date, back to Tierra del Fuego.

FitzRoy asked an acquaintance, the Reverend J. L. Harris, vicar at nearby Plymstock, Devon, to write to the Church Missionary Society on his behalf, for help in placing the Fuegians, at FitzRoy's expense, with some godly folk who might instruct and enlighten them for two or three years. Perhaps he told Harris too much: "They are Cannibals but now they show a ready appetite for Vegetables," wrote Harris. The Society, which was primarily concerned with its missions to Africa and the far east, replied that it did not feel the Fuegians to be within its "province."

In the meantime, the *Beagle* was stripped and cleaned out, and on October 27, its pendant was struck—decommissioned, laid up "in ordinary" once more to await its next call to naval service. The *Beagle*'s crew were paid off and dispersed. "I much regretted the separation from my tried and esteemed shipmates," wrote FitzRoy. "Of those who had passed so many rough hours together, but few were likely to meet again." The *Beagle*'s crew had been a happy one since FitzRoy's appointment. They had spent two years in close company together aboard the tight quarters of the ship, and rowing and camping in the boats for weeks they had shared food and razors and storms, and when inevitably FitzRoy obtained another commission and another ship, he would have to find, train, and mold another crew to his liking again. But for now, his seamen were laid off, equipped with their meager backpay and perhaps a letter of recommendation from their captain, to look for berths aboard other vessels, or to go home to their families, to the farms or cities they came from, to lose or stay in touch with their former shipmates. FitzRoy retained his coxswain, James Bennett, probably on half-pay, leaving him to keep an eye on the Fuegians while he remained in Plymouth to attend to the details of his survey, and to continue to look for the right berth for his cannibal savages.

In early November, he received bad news from Bennett: Boat Memory appeared to have smallpox. FitzRoy contacted Dr. Armstrong at the Royal Naval Hospital in Plymouth, who suggested that all four Fuegians be immediately admitted to the hospital, where they could be isolated and given the best treatment. The Admiralty gave its permission, and the Fuegians were taken there on November 7 and placed in the care of two renowned naval physicians, Drs. David Dickson and Sir James Gordon.

Work on his surveys now took FitzRoy to London for consultation with the Admiralty's hydrographic department, but he had hardly reached there when he received a letter from Dr. Dickson telling him that Boat Memory had died. FitzRoy was filled with genuine grief and remorse.

> This poor fellow was a very great favourite with all who knew him, as well as with myself. He had a good disposition, very good abilities, and though born a savage, had a pleasing, intelligent

appearance. He was quite an exception to the general character of the Fuegians, having good features and a well-proportioned frame. . . . This was a severe blow to me, for I was deeply sensible of the responsibility which had been incurred; and, however unintentionally, could not but feel how much I was implicated in shortening his existence.

On the long voyage home to England, he had gained a good idea of the personalities and abilities of the Fuegians. By the end of it he'd found York Minster "a displeasing specimen of uncivilized human nature." Part of this conclusion was undoubtedly the result of FitzRoy's great belief in the telling aspects of physical appearance. His own portrait drawings of the Fuegians show York Minster, despite a high collar, tie, and frock coat, to be much coarser-featured than the others. His admiration for Boat Memory's appearance was certainly one reason FitzRoy had placed on him his greatest hopes for a civilizing transformation.

Now his concern, and his hopes, centered on the other three, two of them children. They had all been revaccinated on first entering the hospital, and the Navy doctors wrote to FitzRoy that they were optimistic of their chances of resisting the disease.

They didn't get smallpox. But Dickson, in a wild and cavalier act of medical adventuring—and in the best tradition of progressive experimentation; the two inescapably went together—took Fuegia Basket home with him, where his own children had come down with measles. He thought it an excellent opportunity "to carry the little Fuegian girl through that malady" in order to boost her immune system. He only informed FitzRoy later, telling him that he'd prepared her for it and that she'd had "a very favorable attack."

The Fuegians otherwise remained in the naval hospital for the rest of November, while FitzRoy, back in London working on his surveys, anxiously wondered what he would do with them when they were discharged. The rejection by the Church Missionary Society had stalled his plan for their improvement and education.

But suddenly everything got better. FitzRoy's hopes for his Fuegians, naive and unplanned, made him some friends. The Church Missionary Society had not felt able to help him, but its secretary, Dandeson Coates, took a personal interest and put him in touch with the Reverend Joseph Wigram, secretary of the National Society for Providing the Education of the Poor in the Principles of the Established Church. Wigram, then an assistant preacher and later to become the Bishop of Rochester, was the son of a well-to-do landowner and still lived at his father's home in Walthamstow, Essex, just east of London. Gripped with the fervor of his society's avowed mission, he approached Walthamstow's rector, William Wilson, who grew equally excited. Together they hatched a plan for FitzRoy's savages.

8

Early in December 1830, Coxswain Bennett and the *Beagle's* recent master, Mr. Murray, a man now much experienced in his dealings with Fuegians, collected FitzRoy's three charges from the naval hospital in Plymouth. The five of them traveled to London in a privately hired stagecoach. The jolting trip took more than twenty-four hours over rough roads, but Murray told FitzRoy that the Fuegians "seemed to enjoy their journey . . . and were very much struck by the repeated changing of horses."

They were bound for their new home. Rector Wilson and Joseph Wigram had secured a berth for the Fuegians at Walthamstow, where they would be boarded and educated, at FitzRoy's expense, at the local infant school. The schoolmaster would look after and take charge of them. FitzRoy was hugely relieved. He met them with his own carriage at the coach office in Piccadilly and drove them through London to Walthamstow.

Passing Charing Cross, there was a start and exclamation of astonishment from York. "Look!" he said, fixing his eyes on the lion upon Northumberland House, which he certainly thought alive, and walking there. I never saw him show such sudden emotion at any other time.

There was more to see than the statue of a lion. Past Charing Cross, FitzRoy steered his carriage east along the Strand, leaving Mayfair and the tony West End behind, passing through the City of London, the square-mile enclave that by the early nineteenth century had become the financial and business center of the world. Here the streets were lined with banks, shop windows, swinging painted signs, and public houses and filled with a dense circus of costermongers, shoe-blacks, boardmen's advertisements, two-wheeler cabs, and a thronging mob of people. The City was an ancient, squalid, jumble of activity, bordered by the Thames, and the teeming markets of Billingsgate and Smithfield, both of which had been in place for a thousand years.

Only 200 yards from St. Paul's Cathedral, Smithfield was a charnel house nightmare: six acres of slaughterhouses, knackers yards, bone houses, bladder blowers, streets filled with animals alive and hacked to pieces, piles of excrement and steaming entrails, the gutters running with blood and gore. Charles Dickens described Smithfield as a "shameful place . . . asmear with filth and fat and blood and foam." In *Oliver Twist*, he described a typical market morning.

> The ground was covered nearly ankle-deep with filth and mire; and a thick steam perpetually rising from the reeking bodies of the cattle, and mingling with the fog. . . . Countrymen, butchers, drovers, hawkers, boys, thieves, idlers, and vagabonds of every low grade, were mingled together in a dense mass . . . the barking of dogs, the bellowing and plunging of beasts, the bleating of sheep, and the grunting and squeaking of pigs . . . the shouts, the oaths and quarelling on all sides, the ringing of bells, and the roar of voices that issued from every public house; the crowding, pushing, driving, beating, whooping and yelling; the hideous and discordant din that resounded from every corner of the market; and the unwashed, unshaven, squalid, and dirty figures constantly running to and fro, and bursting in and out of the throng, rendered it a stunning and bewildering scene which quite confused the senses.

The same hubbub arose over the selling of fish at nearby Billingsgate, where the streets were awash with a briny soup of fish scales, guts, and blood. And on the other side of the road, along the banks of the Thames, boys and women smoking short clay pipes picked through piles of refuse, looking for tin, old shoes, bones and oyster shells, anything useful, "with faces and upper extremities begrimed with black filth, and surrounded by and breathing a foul, moist, hot air, surcharged with the gaseous emanations of disintegrating organic compounds," as a concerned medical officer wrote.

The smell that washed over the Fuegians as they passed through London was a rich one. The Thames, viscous with sewage and refuse, stank at high tide and low; the streets reeked with decomposing food and litter, "street mud" and "night soil." This blended with "the Scents that arose from Mundung as Tobacco, Sweaty Toes, Dirty Shirts, the Shit-tub, stinking Breaths and uncleanly carcasses." There was the stench of tanneries and glue manufactories, lime kilns, tallow, stables, decaying wood, boiled cabbage, and the odor of graveyards.

From a suburb, the noise of London was said to be "like the swell of the sea-surge beating upon a pebbly shore when it is heard far inland." But in the city the noise was not so white, not the blended, homogenized din of our own time. Contained in the intimacy of narrow streets and alleyways, it was a bedlam of particulate and identifiable sounds: from herds of market-bound animals, roosters and barking dogs, the shouting and shrieking of men, women, and boys, including the incessant hawking of street sellers, the whinnying, rearing, and sharp hoof-fall of horses, the jangle of harness, the thundering of coaches, carts, wagons, ships' horns on the river, the ringing of hand bells and clock towers, the streetworks and traffic bottlenecks that were a constant feature of nineteenth-century London, the hammering of blacksmiths, artisans, coopers, and armorers, the crying of babies.

If the English weather was "fair" that morning in December, it would have resembled at best conditions in Tierra del Fuego. But the climate of London in the early nineteenth century was colder than it is now, and the city was plagued by terrible smogs. Dense cold fogs were made much worse by the widespread burning of "sea coal," a cheap, poor-quality coal carried down the North Sea by coaster colliers from the north of England. Burned in thousands of fireplaces, industrial furnaces, and coal-powered steam engines, sea coal spewed a heavy smoke into the air that condensed over the city like the fallout from a volcano. The poet William Wordsworth and his wife and a friend were forced to abandon their coach a mile from home one day in December 1817 in a fog so dense that the coachman could no longer find his way. They could not see the houses on either side of the street and groped the rest of the way home between railing and curb like blind people, in a fog "not only thick but of a yellow colour [that] makes one as dirty as smoke."

All this and more assaulted the Fuegians, who had just spent a month in the relative quiet and fresh air of Plymouth. Twenty-first-century humans, reared on *Star Trek* and *Star Wars*, abducted and taken to another world by intergalactic travelers, would probably be less vertiginously displaced than the natives of Tierra del Fuego in London.

Their only comfort in this maelstrom of dislocation were their shipmates, Murray, Bennett, and preeminently FitzRoy, the evident master of this universe. They must have placed in him a growing trust born of overwhelming need.

But a few miles farther on, the city fell away to countryside. Only seven miles from London, as it existed then, Walthamstow lay beyond open fields and across the river Lea, at the edge of Epping Forest, a small country town of 4000 inhabitants. The only way to reach it in 1830 was by road; the railway would not connect with it until the 1840s.

The master of the infants school, William Jenkins, and his wife awaited them. FitzRoy recorded the arrival.

[The Fuegians] were much pleased with the rooms prepared for them at Walthamstow; and the schoolmaster and his wife were equally pleased to find the future inmates of their house very well disposed, quiet, and cleanly people; instead of fierce and dirty savages. . . . The attention of their instructor was directed to teaching them English, and the plainer truths of Christianity, as the first object; and the use of common tools, a slight acquaintance with husbandry, gardening, and mechanism, as the second.

The Walthamstow Infants School had been established by the Reverend William Wilson in 1824. Wilson had come under the sway of a well-known educator of the day, Samuel Wilderspin, who believed that the age level for children entering national schools—six and seven—was far too late. Wilderspin thought that two was the preferred age, particularly for instruction in spiritual matters, which he felt were too important to be left to the haphazard example of poorly educated parents of dubious moral values. Wilson, the wealthy son of a manufacturer, was persuaded by Wilderspin to sponsor the country's first Church of England–sanctioned infants school. It was initally housed in a barn. A few years later Wilson had a new building constructed for the school beside the church graveyard.

Here the Fuegians sat learning English, arithmetic, and "the plainer truths of Christianity" in a classroom adorned with biblical pictures and quotations.

In 1825, Wilson published a book, *The System of Infants Schools*, in which he described his ideas and methods. He wanted to see children at school as early as possible,

[in] the most impressible years of our existence. The evil which is within us is then fomented, or the principles of religion and moral excellence were then first inculcated and encouraged. . . . Muscular action is made a component and necessary part of the system. Every lesson is accompanied with some movement of the person . . . the whole frame is at different periods called into action and restored to rest. The beat of the foot, the clap of the

hands, the extension of the arms, with various other postures, are measures of the utterance of the lesson as they proceed. The position is also frequently changed. The infants learn sitting, standing or walking.

Fuegia, aged nine or ten when she arrived at Walthamstow, and Jemmy, about fourteen, made good progress. Their eight months of captivity aboard the *Beagle* had equipped them with a rudimentary vocabulary—a sailorly argot hardly suitable for a Christian infants school but which nonetheless readied them for further instruction. By all accounts, the two younger Fuegians were responsive, eager pupils, and their "improvement" charmed their benefactors.

They exhibited distinct traits, recorded by FitzRoy and others: Jemmy Button showed a taste for English dress that bordered on dandyism. He "was fond of admiring himself in a looking glass." His speech became peppered with some of the quainter expressions of the day, so that his few quotes that have come down to us (probably because his listeners were amused by them) seem almost to satirize his dress and situation: "Hearty, sir, never better," he would respond to an inquiry after his health. He must have been fun; the butt of much humor, but enjoying the joke himself, and happy to provide it. He made many friends.

Little, round Fuegia Basket exuded an empathic sweetness that endeared her to everybody. FitzRoy had been the first to come under her spell, and had been more unwilling to part with her, to send her ashore with the other child hostages in Tierra del Fuego, than he was able to convincingly explain. "She seemed to be so happy and healthy, that I determined to detain her as a hostage for the stolen boat," he wrote at a point when such hostage-taking had proved futile. To the *Beagle*'s crew, who had made her "a pet on the lower deck"; to the English families she stayed with in Montevideo and Rio de Janeiro, and to the many people she met in England, Fuegia was an irresistible charmer.

But the hulking, brooding York Minster, aged twenty-six, charmed no one. He did not enjoy the classroom, in which he was expected to sit alongside two-, three-, and four-year-olds, singing, clapping, mouthing by rote his ABCs and chants about how to wash one's face and hands, exercising his man's body in tandem with the movements of babes. There would have been no cultural gap large enough to save him from the most abject humiliation in this. He preferred mechanical instruction in smith's or carpenter's work; he paid attention to what he saw and heard about animals; he reluctantly helped with the gardening around the school and church, perhaps feeling this was women's work; but it's not surprising that he had "a great dislike" for schoolwork. Or that he was moody and unresponsive to the finer influences he was exposed to. York was too old and fully formed to change, to sparkle with Christian values. He remained a true captive, imprisoned in himself by a culture that had no place or use for him, no appreciation of his natural self, that wanted to eradicate the skills and tendencies formed in him by his own, and only natural, environment. The difference in age between him and his fellow Fuegians was too great to allow for any kind of real friendship with them. He was as marooned in England as was Crusoe on his island, and though no one seems to have guessed as much, blaming his sullen behavior on savage intractability, he must have been a profoundly lonely and unhappy man.

Because he failed to charm FitzRoy, York Minster was possibly not given the chance to charm anyone else. Fuegia Basket and Jemmy Button, because of their youth and more malleable personalities, were taken on outings with FitzRoy. Whether York Minster accompanied them is unclear: "They [the Fuegians] gave no particular trouble; were very healthy; and the two younger ones became great favourites wherever they were known. Sometimes I took them with me to see a friend or relation of my own, who was anxious to question them," FitzRoy wrote. If York was left behind on these occasions, it wouldn't have helped his attitude.

But the other two became sought-after guests. FitzRoy's family, social, scientific, and professional worlds gave him an extraordinarily broad acquaintance across the dominant strata of English society. Everyone who knew him or heard about FitzRoy and his savages would have been curious to see them, and he received on their behalf more invitations than they were able, or than he saw fit, to accept. They visited wealthy aristocrats, the foundation of FitzRoy's world, who received them in large houses dazzling with furnishings and oversize paintings, serviced by retinues of servants. They were frequent visitors to FitzRoy's sister, Fanny Rice-Trevor, whom the Fuegians called "Cappen Sisser." She gave them many gifts, paid great attention to them, took them shopping for clothes with FitzRoy or coxswain Bennett, and they developed a real affection for her, talking of her often at the time, and long afterward. Fanny moved in exalted circles, attending events at court such as the queen's birthday, and undoubtedly spoke of her brother's "Indians" to many people who would have been eager to meet them.

Other hosts would have been FitzRoy's acquaintances in the scientific and professional world, such as Roderick Murchison, a foremost geologist who later became president of the Royal Geographic Society and remained keenly interested in the Fuegians' welfare for years after meeting them; Cambridge professor-geologist Adam Sedgwick, one of the most influential educators of his day; probably Charles Lyell, the young barrister-turned-geologist, whose book *Principles of Geology* enormously influenced FitzRoy's (and everyone else's) thinking about the age of the world and its natural history; and his navy friends: Francis Beaufort, the hydrographer of the British Navy who later developed the Beaufort wind scale; Sir John Richardson, the retired naval expert on the natural history of the Arctic; and many others who wished to see these examples of "brute Creation" whom FitzRoy had plucked from the wild and given a swift makeover as English gentlefolk.

On these occasions, FitzRoy's specimens would be encased in their stiff, Sunday-best clothes, rather than their schoolday garb, and would arrive at their hosts' homes with him in his carriage, drilled in the basic graces and polite exchanges. They were questioned about their native land and about their present circumstances; they were served refreshments and full meals. They were given presents. Their answers, in broken English, and their developing table manners, invariably charmed and fascinated their hosts.

The culminating pinnacle of the Fuegians' social forays was an audience with the new king, William IV, and his wife, Queen Adelaide. A quiet man and lackluster monarch with a disdain for pomp and ceremony, William had been welcomed by the British public and its government on succeeding his brother, George IV, whose tabloid escapades had been an embarrassment to the country. Never expecting to become king, William had lived quietly with his mistress, Mrs. Dorothea Jordan, for twenty years and fathered ten illegitimate children by her. None of these was a possible heir, so on accession to the throne in 1830 he married Adelaide of Saxe-Coburg and Meinengein, who bore him two daughters, both of whom died in early childhood.* William had joined the Royal Navy at the age of thirteen, traveled widely, and was known as the Sailor King; he hungered for associations beyond the court, for his earlier connection with the exciting world of travel. As king, he often invited explorers and adventurers to meet with him and questioned them about their exploits. With his naval connections and interest, he clearly knew about Captain FitzRoy.

The only record of the visit is FitzRoy's.

During the summer of 1831, His late Majesty expressed a wish to see the Fuegians, and they were taken to St James's. His Majesty asked a great deal about their country, as well as themselves; and I hope I may be permitted to remark that, during an equal space of time, no person ever asked me so many sensible

---

*William's niece, Victoria, succeeded him on his death in 1837.

and thoroughly pertinent questions respecting the Fuegians and their country also relating to the survey in which I had myself been engaged, as did His Majesty. Her Majesty Queen Adelaide also honoured the Fuegians by her presence, and by acts of genuine kindness which they could appreciate, and never forgot.

It is Fuegia Basket who is singled out for mention in FitzRoy's account of the meeting. The sweet natural empathy that had prevented him from giving her up in Tierra del Fuego drew Adelaide's greatest attention. The childless queen was captivated by the gamine native girl in her Christian frock.

> She left the room, in which they were, for a minute, and returned with one of her own bonnets, which she put upon the girl's head. Her Majesty then put one of her rings upon the girl's finger, and gave her a sum of money to buy an outfit of clothes when she should leave England to return to her own country.

As an accomplished amateur scientist and already a renowned explorer, FitzRoy was exhibiting "his" Fuegians as performing curiosities, just as Professor Challenger would unveil his leathery pterodactyl to the members of the Zoological Institute at the conclusion of Arthur Conan Doyle's *The Lost World*. They were the fruit of the obsession he had picked up, like sea fever, in Tierra del Fuego, though his interest in them for their own sakes, and for their welfare, was altruistic and sincere. But he felt more than a professional glow of pride at their accomplishments. The space he gave to them in the *Narrative*, and the language of concern for their well-being and happiness, indicate a strong emotional involvement. FitzRoy was still a young man, only twenty-five in that winter of 1830–1831 when he stepped out in English society with his charges, unmarried and living alone, except for his servants. He almost certainly felt for the two younger Fuegians, children in his care, a fatherly emotion. Perhaps even love.

• • •

When not squiring his Fuegians about on their social engage-
ments, FitzRoy spent the winter and first three months of 1831
in the Hydrographic Office of the Admiralty supervising the
drawing of charts based on his surveys of Tierra del Fuego, and
writing the sailing directions to accompany them. The vast
labyrinth of southern South America still remained largely unexplored, yet FitzRoy, Captain Phillip Parker King, and the late
Captain Stokes and his more functioning officer Lieutenant
Skyring had achieved a great deal for vessels navigating between
the Atlantic and the Pacific.

> Two escape routes from the southern part of the Magellan Strait
> direct to the Pacific Ocean had been charted, enabling west-
> bound vessels encountering northwest winds to make their way
> more quickly to the open sea; an inshore route had been found
> on the west coast from the Gulf of Peñas direct to the Magellan
> Strait for small vessels using the prevailing northwest winds; the
> rugged coasts around Cape Horn had been charted, as had the
> Strait of Lemaire; and the remote and dangerous Diego Ramirez
> Islands had been fixed where they lay, far out to the southwest
> of Cape Horn. (*The Admiralty Chart*, Rear Admiral G. S.
> Ritchie)

British Admiralty charts had been made available for sale to
the merchant fleets of the world since 1821, and the new charts
and sailing directions that were the fruit of the *Beagle*'s and the
Adventure's commission would have been of invaluable assistance to mariners and navigators for the cost of a few shillings.

Four draughtsmen worked in the Hydrographic Office. These
men drew the charts from FitzRoy's and King's drawings, under
their supervision. When the drawings were completed they were
taken to Messrs Walker of Castle Street, Holborn, for engraving
onto copper plates. The plates were delivered back to the Admi-

ralty, where they were used on the navy's copper press, by its own copper printer and assistant, whenever a run of charts was required.

FitzRoy was doing this work at a time of signal change in the Hydrographic Office, under the direction of a figure whose name is known to seamen the world over today. Two years earlier, Captain Francis Beaufort had been appointed Hydrographer of the Navy (the choice had been between Beaufort and Captain Peter Heywood, the last survivor from the *Bounty* mutiny). When Beaufort took over the chair in May 1829, the Napoleonic wars had been over for fourteen years, and Britain's navy was in the early days of a century of peace in Europe that would last (apart from the interruption of the Crimean War) until the outbreak of World War I. The navy's interest had shifted from defense to the guardianship of its empire, and facilitating the expansion of trade and exploration. Beaufort intensified surveying efforts across the globe, and within a few years of his appointment sent ships and surveyors back to South America; to Africa, Australia, New Zealand and New Guinea, the South China Sea, the Caribbean, Canada, the Mediterranean; and to the home waters around the British Isles. A number of these surveyor-captains who became famous for their work—Skyring, Lort Stokes, and Sulivan—got their training with FitzRoy aboard the *Beagle*.

Beaufort was fifty-five when he became hydrographer. His interest in charting came from his father, the Rector of Navan, County Meath, Ireland, an amateur cartographer who had made an excellent map of Ireland. Beaufort joined the navy at thirteen and served under several surveying commanders, including Alexander Dalrymple, who had prepared a chart of descriptions of wind strengths, numbered 0 ("flat calm") to 12 ("a hurricane such that no canvas could withstand"). Beaufort refined Dalrymple's scale to give wind speeds in nautical miles per hour and accompanying descriptions of sea states for each "force" on the scale (for example, "Force 8; 34–40 n.m.p.h; Near Gale; Height of Sea in ft: 18; Deep sea criteria: High waves of increasing length, crests form spindrift"). He persuaded the navy's captains and navigators to employ it in descriptions of sea conditions in their logs and official reports, and today the Beaufort Scale is used and understood by sailors the world over.

Like FitzRoy, Beaufort had scientific interests that took him far beyond the navy and made him useful friendships. He was a Fellow of the Geological Society, the Royal Society, the Astronomical Society, and a leading figure in the founding of the Royal Geographical Society in 1830, just as FitzRoy was returning from Tierra del Fuego. The rapid efflorescence of science in the early nineteenth century led the Admiralty, with Beaufort's urging, to form its own scientific branch in 1831, to contain the Hydrographic Office, the Royal Observatories at Greenwich and Cape Town, the Nautical Almanac and Chronometer Offices, and later the Compass Office. Beaufort was the ideal link between the musty, wooden world of the Georgian navy and the progressive, expanding, enlightened Victorian era that was about to take over.

A sudden and important presence in the Hydrographic Office over the winter and early spring of 1830–1831, FitzRoy met with his superior regularly. Half Beaufort's age and always fully conversant with the latest scientific developments, FitzRoy not surprisingly became one of the hydrographer's favorites.

Lucky for him. He was soon to have urgent need of Beaufort's help.

As the two Fuegian children continued to thrive at Walthamstow, their adult compatriot sank deeper into isolation. Cut off from any meaningful contact with the world, he was also cut off as a man. FitzRoy estimated York Minster's age as twenty-six in 1830. He was in the physical prime of his life. At home, among his own people, he might have had (possibly he did have) a wife. Certainly he would have been having sex. The straitlaced confines of the church-run Infant School, the constant supervision of Schoolmaster Jenkins, his wife, the Reverend Wilson, and

Coxswain Bennett, whom FitzRoy had installed in Walthamstow to keep him posted on the Fuegians, and the unceasing importuning to strict Christian behavior must have been a living hell to a healthy, primitive man. The flourishing prostitution industry in London was far beyond his reach, geographically and socially.

There is no record of any associations that might possibly have occurred between York Minster and the maidservants and women he would certainly have come into contact with at Walthamstow. There is only the known fact that at some point during their stay there, he fastened his sexual attentions on the ten-year-old Fuegia Basket.

9

It was FitzRoy's man on the scene, Coxswain Bennett, installed in rooms nearby, whose worldly experience beyond straitlaced Walthamstow made him the best and earliest witness to what was afoot.

And so it went until late one afternoon when Coxswain Bennett happened down the hedge by the Infants School. Fortunately he had left his sharp cutlass at the inn. But, shifting his quid, spitting to windward, and with a look of determination on his mahogany face, he grabbed Fuegia, as York disappeared through the hedge and, turning her over his honest knee, spanked her soundly.

"It no me, no me! *York, he York, only York,*" she pleaded.

"And so, sir," reported the coxswain, "I quits aspankin' of her fat little bare bottom, sir, and sets her down . . . but the way that York's been aglarin' at me since has me awearin' of my cutlass, an' I sleeps with it, sir, when I does sleep."

"Bennett, this is the most terrible thing that yet has happened. I'd rather—I'd rather see her in a Christian grave, like poor Boat." Captain FitzRoy, in the snug at the inn, to which he had hastened by private appointment, broke a life's rule. He ordered port, and he and his coxswain fortified themselves in private.

"Them heathen women ripens rapid, sir. I mind me, in Calicut."

"Yes, Bennett—yes, I know. But what *is* to be done?"

"Sail, sir, sez I. Slip the cable, Captain, an' I'm with you, sir."

This account comes from *Cape Horn*, by Felix Reisenberg, a generally splendid, thoroughly researched history of Cape Horn and the mariners and natives who have made the place famous. Reisenberg, a seaman of wide experience, a Columbia graduate, engineer, maritime historian and writer, had a novelist's feel for situation and character and relates history with gusto. But he was writing a book for the popular market and had no qualms about taking some liberties. This episode, suggesting that Coxswain Bennett caught York Minster and Fuegia Basket *in flagrante delicto* down by the hedge at the Infants School, is clearly one of them. Fanciful, unsupported, and written as if Reisenberg had stayed up too late reading *Treasure Island*, it's the suggestive missing link in the story of FitzRoy's Fuegians. It is what can be inferred from what was known and subsequently happened, but is nowhere reliably recorded.

Most discordant is Reisenberg's FitzRoy. The captain wringing his hands and asking his coxswain "What *is* to be done?" sounds wholly out of character for the imperious young FitzRoy. Maddeningly, tantalizingly, Reisenberg frequently "quotes" the garrulous coxswain, making of him a salty and salacious commentator on great doings that have been handed down and sealed in the often dull and cloudy aspic of nineteenth-century prose.

But during the spring of 1831, *something* happened between the Fuegians at Walthamstow to make FitzRoy suddenly and drastically curtail his original plans to educate them "and, after two or three years . . . take them back to their country." Because by June, after the Fuegians had been in Walthamstow only seven months, FitzRoy was going to extraordinary lengths, and digging deep into his own pockets, to get them out of England and back to Tierra del Fuego as soon as possible.

His official work supervising the production of charts from his surveys was finished in March. "From the various conversations which I had with Captain King . . . I had been led to suppose that the survey of the southern coasts of South America would be continued; and to some ship, ordered upon such a service, I had looked for an opportunity of restoring the Fuegians to their native land." But there was no immediate prospect for a further South American commission. Despite Beaufort's personal desire that the survey of South American waters be continued, and FitzRoy's accomplishments there, the Admiralty considered an early return to Tierra del Fuego unnecessary in the spring of 1831. Yet, certainly within the two- to three-year time frame of FitzRoy's original plan for the Fuegians, he would have found a ship bound around Cape Horn that could have taken his passengers. But now he "became anxious about the Fuegians." And, after so closely detailing his hopes for them, that is all he writes about the sudden turnaround of his scheme.

So anxious that he arranged to charter a small merchant vessel, the *John of London*, to carry him, Coxswain Bennett, the three Fuegians, and the supplies amassed and donated to them back to Tierra del Fuego, and eventually land him and Bennett in Valparaiso, where they could find a ship returning to England. No little jaunt this, the cost of chartering the *John* was £1,000—the equivalent of the price of a London townhouse—to which FitzRoy would have added food and supplies for himself and his retinue for at least six months, plus salary for Bennett.

York Minster's attentions to Fuegia Basket were almost certainly responsible for such an extreme departure from his initial design. FitzRoy later admitted as much.

> He had long shown himself attached to her, and had gradually become excessively jealous of her goodwill. If anyone spoke to her, he watched every word; if he was not sitting by her side, he grumbled sulkily . . . if he was . . . separated . . . his behaviour became sullen and morose.

An episode like the one imagined by Felix Reisenberg, or the slightest suggestion of it, would have been disastrous for FitzRoy. The unseemly possibility that Fuegia Basket might become pregnant was the least of it. This indication of ineradicable savagery breaking through the civilizing glaze on his charges threatened to destroy all the goodwill that had been showered on the Fuegians and FitzRoy himself by the royal family, the court, the Admiralty, FitzRoy's friends, the public at large, Reverend Wilson, and the kind people at Walthamstow. All had embraced the Fuegians and taken deeply to heart FitzRoy's own pious sentiment that they offered a rare and quite clearly divinely engineered opportunity not only to deliver three heathen souls from the perils of savagery but to return them home to plant a holy seed that might grow and spread across a Godless continent. Reaction to news that the sweet child adored by so many was having sex with a hulking, unreformable savage at the Walthamstow Infants School would be catastrophic. It could mean disgrace for FitzRoy and the end of his naval career.

On a more personal level, this development deeply shook FitzRoy's faith in his grand scheme. He was so certain of the benefits of exposure to the Anglo-Christian civilization that he saw physiological proof of it: he believed the Fuegians' physical features were altered and improved by their stay in England.

The nose is always narrow between the eyes, and, except in a few curious instances, is hollow, in profile outline, or almost flat. The mouth is coarsely formed (I speak of them in their savage state, and not of those who were in England, whose features were much improved by altered habits, and by education).

FitzRoy's own ink portraits of Jemmy Button reflect this clearly. A comparison between Jemmy in his "savage" state, and the "English" Jemmy, wearing a high collar, cravat, and frock coat, shows features altered as if by cosmetic surgery: Anglo Jemmy's nose and mouth are finer, flared and expressive, the

FUEGIA BASKET. 1833.

JEMMY'S WIFE. 1834.

JEMMY IN 1834.

JEMMY BUTTON IN 1833.

YORK MINSTER IN 1833.

YORK IN 1833.

FUEGIANS.

eyes bright with acuity as if he were attending a lecture on paleontology, his posture upright and proud. Savage Jemmy looks a
million years older, a Neanderthal version with an entirely different cranium shape, crouched and hulking, thicker-featured, with
dull-witted eyes sunk beneath a heavier brow.

The most fascinating aspect of these two portraits is that the
Anglo Jemmy is dated 1833, the Savage, 1834, a year *after*
Jemmy Button had been returned to Tierra del Fuego. FitzRoy
chose to believe that such physical improvements lasted only as
long as the heathen remained under the beneficent sway of God-
fearing English culture. Once back in the wild, Jemmy's lips
thickened, his nose widened, his brow bulged, his cranium
reverted to its primordial shape.

FitzRoy's portait of York Minster in England shows a largely
unimproved, thick-featured Fuegian with a tie and a haircut.

There was an immediate change in their living arrangements.
Although FitzRoy later wrote that the Fuegians remained in
Walthamstow until October of 1831, this was not the case. In
August, the Reverend Wilson referred in a letter to their "late resi-
dence" there: FitzRoy had removed them. Where he put them, and
to what degree Fuegia and York were separated, is not known.

But his urgency to leave England with them was real, and finally
brought a change at the Admiralty. He confided his anxiety, and his
plans to charter the *John,* to an uncle, the Duke of Grafton, who
then interceded on his behalf. The duke, together with Beaufort,
persuaded the Admiralty lords that FitzRoy's earlier survey

FitzRoy's Fuegians, from his own drawings. *Top left*: the gamine Fuegia
Basket in England, aged 10. *Center right*: Jemmy Button around 1832,
after several years of civilizing influence. *Center left*: Jemmy Button in
1834—FitzRoy believed even his cranium and features had improved in
England, and then regressed again to primitive shape after just a year
back in Tierra del Fuego. *Bottom*: York Minster as FitzRoy saw him at
every stage, an intractable savage. *Top right*: Jemmy's wife in 1824.
(Narrative of HMS *Adventure* and *Beagle*, by Robert FitzRoy)

remained seriously incomplete: trade restrictions between Britain and the newly independent confederation of Latin American states had recently been lifted; French and American interests in the area rivaled and threatened to displace any possible British presence; a more complete survey of southern South American waters, plus the possibility of installing a group of friendly natives, perhaps under the auspices of the Church Missionary Society, could prove to be an incalculable political and economic opportunity.

Very quickly, FitzRoy was commissioned to embark on a new survey. He still had to pay most of the £1000 he had promised for the *John*, but now he had what he wanted: a chance to continue the work he had started, which was demanding and provided the best peacetime opportunity for him to shine as a naval officer—and the means to cut and run with the Fuegians.

Initially, he was appointed to the command of the *Chanticleer*, a near sistership to the *Beagle,* recently returned from a surveying voyage in South American waters, where her commander, Henry Foster, the Admiralty's leading field astronomer, had drowned in an accident on the Chagres River earlier in the year. But when examined, the *Chanticleer* was found to be too tired for such a voyage, her planking sprung, her gear and rigging worn out from her punishing travels in the Southern Ocean and the tropics. Another ship was quickly found; to FitzRoy's delight, it was the *Beagle*. He threw himself into preparations.

The *Beagle* was commissioned anew on July 4, 1831, and immediately given prime dock space at Devonport, and the work needed for a long and arduous period at sea commenced. With FitzRoy at the helm, Beaufort devised a much grander plan for this second voyage of the *Beagle*. In addition to the renewed survey of South American waters, he proposed that FitzRoy return to England by sailing westabout around the world, across the Pacific, through Australasia, across the China Sea and the Indian Ocean. This would enable FitzRoy to track an unbroken chain of meridian distances around the world, which, depending on the accuracy of the chronometers carried with him, would enable the precise charting of the longitude of every port the *Beagle* called at. Much of the charting and surveying of the Pacific remained unimproved since Captain Cook's time, and this would be a significant opportunity to refine the mapping of the world, on which hung the expansionist improvements of trade, political power, and colonial possession. In addition, Beaufort and FitzRoy discussed a wide variety of botanical, geological, and meteorological inquiries that might be pursued on such a voyage, turning this circumnavigation into a showpiece of scientific endeavor. FitzRoy, the scientist-captain, was thrilled: "I resolved to spare neither expense nor trouble in making our little expedition as complete, with respect to material and preparation, as my means and exertions would allow, when supported by the considerate and satisfactory arrangements of the Admiralty."

But beneath the exciting prospects, he also felt a gnawing concern. It would be a long voyage, several years at least, possibly three or four, during the whole of which he would be effectively alone, cut off from the world and even, especially, the men aboard his ship. FitzRoy was not a friendly commander, not an open-hearted Jack Aubrey or a matey Peter Blake. Although he had sailed and rowed and shared every hardship and success with his officers and men, he lacked ease with them. He was an aristocrat, high-strung, given to fits of depression and querulous anger. There would be no escape for him from the loneliness and isolation of command. This had driven Pringle Stokes, in whose cabin he would now spend years more, to shoot himself. There was madness in FitzRoy's own family which had resulted in the suicide of his uncle.

He might have paid less heed to this had it not been for the essential failure of his Fuegian enterprise. This had crucially undermined his confidence in his judgment; he found himself, for the first time in a brilliant career, faltering, unsure. He was suddenly afraid of being so alone.

In August, as work progressed aboard the *Beagle*, he approached Beaufort with an unusual request.

> Anxious that no opportunity of collecting useful information, during the voyage, should be lost; I proposed to the hydrographer that some well-educated and scientific person should be sought for who would willingly share such accommodations as I had to offer, in order to profit by the opportunity of visiting distant countries yet little known.

FitzRoy was asking if he might take a companion.

Beaufort agreed, even thought it a good idea, and wrote to a friend, George Peacock, a mathematics professor and fellow of Trinity College, Cambridge, asking him to recommend some "savant" to serve as a naturalist aboard the *Beagle*.

Were it not for Robert FitzRoy's concern over the sexual tension between Fuegia Basket and York Minster, this second voyage of the *Beagle* would not have occurred at that moment and under the circumstances that it did. Had the timing, or FitzRoy's need, or his new sense of his own fallibility, been any different, the door would never have opened for a Shropshire doctor's son—an undistinguished bachelor of arts graduate who, for want of ambition, was preparing to become a clergyman—to voyage around the world and shatter and remake the way we think of ourselves in the profoundest way.

But he almost didn't go. He was hardly anyone's first choice.

10

Perhaps only the dawn of the Internet, and the computer technology that coalesced with it into a global ethos at the end of the twentieth century, can give any idea of the excitement generated by the sciences of the physical world in the early nineteenth century.

Geology was preeminent among these, for its findings had recently shaken the widely held belief that the earth had existed for a mere few thousand years. Calculations from biblical and historical records had previously indicated that the epic first week of creation, described in the book of Genesis, had begun on October 22, 4004 B.C. Six days later, by October 28, Earth and all its glories, including Man, were in place; and on October 29, God rested.

Science and the Bible had for a time even become comfortable bedfellows. The perennial discovery of fossilized sea creatures far inland seemed to support the biblical story of the great flood that had once washed over the earth.

And God said unto Noah, The end of all flesh is come before me; for the earth is filled with violence through them: and behold, I will destroy them with the earth. Make thee an ark of gopher wood: rooms shalt thou make in the ark, and shalt pitch it within and without with pitch. And this is the fashion which thou shalt make it of: The length of the ark shall be three hundred cubits, the breadth of it fifty cubits, and the height of it thirty cubits. (Genesis 6: 13–15)

The cubit, the biblical unit of measurement, was generally thought to be the length of a man's arm from elbow to the tip of the middle finger. Men differed, but Sir Isaac Newton set the matter at rest by determining that the cubit must be 20½ inches, and was then able to calculate that Noah's ark, at 300 cubits by 50 cubits, had been 537 feet long, 85 feet wide, 51 feet high, and weighed (Newton must have decided on a unit weight for gopher wood, *Cladrastis lutea,* and construction scantlings) 18,231 tons.

This was science at its most tidy and helpful.

But the deeper geologists looked, the more they saw that confounded them: fossils, and the arrangements of layers and layers of earth that had preserved them, began to indicate subterranean upheavals, erosion, sediment, the existence of ancient seas—signs of tremendous change taking place across the face of the earth. Either these changes had occurred at one cataclysmic moment—during the flood, handily explained by the Bible—or, as geologists began to think, they had taken place over an immense period of time, and were still taking place, continually, but with imperceptible slowness, suggesting such a mind-boggling age to the planet that science and the Holy Word could not be reconciled.

The result, for those who did not cling to the literal word of the Bible, was an abyssal unknown, a spiritual vacuum. Science rushed in to fill it, and its discoveries, coming in tumbling profusion in the early years of the nineteenth century, were greeted with the excitement of bulletins from a war front. Nowhere was this zone hotter, of greater moment, more closely watched, than in the circles of scientific inquiry at Britain's universities. In particular, at Cambridge.

Beaufort's letter to Professor Peacock asking about a companion for FitzRoy tapped into an elite cream of intellectual movers and shakers at the very top of the British establishment. Cambridge professors like Peacock, Adam Sedgwick, John Stevens Henslow, William Whewell, and John Herschel were intimates of men in the government and the armed forces, men like Beaufort, the leading scientific light at the Admiralty, and the prime minister, Sir Robert Peel. They were close to mathematicians like Charles Babbage, whose series of increasingly complicated "difference engines" were the world's first computers; to geologists like Charles Lyell and Roderick Murchison. Whewell and Herschel wrote books that became, along with Lyell's *Principles of Geology*, the most read, talked-about, and influential books of their day. Herschel's *A Preliminary Discourse on the Study of Natural Philosophy* inspired and laid the groundwork for much of the scientific induction and explanation that followed its publication in 1831. Whewell's *History of the Inductive Sciences* was acted out in a game of charades at Lord Northampton's Christmas party, achieving the sort of buzz generated by popular television shows a century and a half later. These men created an intellectual aristocracy at the core of the world's greatest empire and shaped the way that world thought.

Beaufort's letter—an invitation to sail away and examine the still largely unknown world aboard a well-stocked floating laboratory—sent a tremor through their community.

Peacock in turn wrote to John Stevens Henslow, professor of botany at Cambridge, who had a wide acquaintance with a number of scientifically inclined young gentlemen who might be suitable for the voyage Beaufort described. Henslow, who had also been a professor of mineralogy, was described as a man "who knew every branch of science." He was a cornucopia of prevailing scientific knowledge, and his lectures were immensely popular and crowded, attended even by other professors. Harder to get into were Henslow's Friday evening soirées, where ten to fifteen favored students and professors could informally discuss the latest and headiest intellectual and scientific ideas. He led field trips, on foot, on horseback, by stagecoach or barge, that might end with supper at an inn or tavern.

Henslow's students were mainly well-to-do upper-middle-class young men who came to school with dogs, guns, and horses and set themselves up in private lodgings around Cambridge, attending lectures and their studies only when these were fun. There were a few budding scientists among them, but most were preparing for roles as doctors, barristers, politicians, landowners, and clergymen in the dominant establishment from which they had sprung, a kind of extended family of plutocracy. They were familiar with the classics in their original Greek and Latin, they felt the ease of entitlement in company, they rode, shot, and drank well. They were gentlemen. One of these, with an enthusiasm for the natural sciences perhaps more developed than in his fellows, was what Beaufort, on FitzRoy's behalf, was looking for.

"What treasures he might bring home with him," Peacock wrote to Henslow, "as the ship would be placed at his disposal, whenever his inquiries made it necessary or desirable. . . . Is there any person whom you could strongly recommend: he must be such a person as would do credit to our recommendation. Do think on this subject: it would be a serious loss to the cause of natural science, if this fine opportunity was lost."

For just a moment, Henslow thought of going himself. He was even then considering a trip with several students to Tenerife, in the Canary Islands, a place that had been described by Alexander von Humboldt as a scientific paradise. As a young man, Henslow had read François Levaillant's *Travels from the Cape of Good Hope into the Interior Parts of Africa* (1790), the story of the Frenchman's shipwreck on the South African coast and his trek through the country with just a rifle, ten gold ducats, and the clothes he'd washed ashore in. Henslow had become gripped by the urge to make the same expedition and day-dreamed of Africa and travel. Now, aged thirty-five, with work, a wife, and a new baby pinning him to a modest house in England, he held the thought of this incredible voyage around the world in his palm for a moment, then ruefully passed it on.

He sent Peacock's letter on to his brother-in-law, Leonard Jenyns, a Cambridge graduate, now a curate at nearby Bottisham, and an amateur entomologist who was highly respected among the Cambridge naturalists. Jenyns too was immediately gripped by the idea of the voyage, enough to begin thinking about what clothes to take. But he had recently been appointed to his curacy and reluctantly concluded that it was not "quite right to quit for a purpose of that kind."

To both Henslow and Jenyns, the voyage seemed self-indulgent, the sort of thing a grown man of responsibility could not seriously consider. It appealed to the boys they had once been but felt they could no longer be. They agreed to send Peacock's letter on to such a boy, a Cambridge student, just graduated, who had charmed them both with his naturalist enthusiasms and who was still, unlike them, on the other side of the threshold of responsible manhood.

Charles Darwin, aged twenty-two, was, in fact, the student who had whipped Henslow up about a trip to Tenerife. It was Darwin who had read von Humboldt's *Personal Narrative* of his journey to Tenerife and through the Brazilian rain forest in 1799–1804. That book had instilled in him a sudden, almost urgent desire to travel, and "to add even the most humble contribution to the noble structure of natural science." Brazil was far away, an expensive and difficult destination for a brief visit by an amateur naturalist. But Tenerife, off the coast of Morocco, was much closer; von Humboldt's descriptions of the island, visited en route to Brazil, had started Darwin talking to Henslow about a summer expedition there.

But a voyage around the world! Darwin decided immediately to accept, but his father just as quickly expressed his deep disapproval.

To Dr. Robert Darwin, the notion of his son suddenly head-

Charles Darwin at 31, shortly after his voyage in the *Beagle*.
(*Watercolor of Charles Darwin by George Richmond;
by permission of English Heritage and Down House.*)

ing off around the world seemed one more sidestep in the pattern of irresolution and disinclination to settle into a profession that Charles had shown since his earliest days at university.

At sixteen, he was doing so poorly in school that his father decided he was wasting his time and sent him a year early to Edinburgh to join his older brother Erasmus in studying medicine to become a doctor. Both boys had been keen "scientists" at home, setting up their "laboratory" in an old garden shed. They bought glass-stoppered bottles and heated to incineration with an Argand lamp coins and whatever else would burn in the fireproof china dishes donated by their uncle, the wealthy potter Josiah Wedgwood. They analyzed minerals, chemicals, and compounds supplied by their local chemist in Shrewsbury. Darwin became fascinated by crystallography and began collecting rocks.

Erasmus's chronic ill health made him a poor candidate for a doctor, and their father fastened his hopes on Charles. He also thought Charles's amiable nature would make for a sympathetic bedside manner. But the young Darwin discovered that he was repelled by the practical side of medicine. Apart from the revulsion he felt for dissection, the trade in bodies used in anatomy classes carried its own notorious associations. The subjects were supposedly the dead from hospitals and the poorhouse, or deceased or executed criminals, but they were frequently victims murdered for the sale of their corpses. In 1828, three years after Darwin arrived there, William Hare and Irishman William Burke killed at least sixteen people in Edinburgh's Old Town and sold the bodies for cash—£10 in winter, £8 in summer, when preservation proved more problematical—at the medical school's back door. Other corpses came from grave robbers and body dealers who had them shipped in barrels of cheap whisky from city slums and across the Irish Sea from Dublin.

Darwin found operations on living subjects even less tolerable.

I . . . attended on two occasions the operating theatre in the hospital at Edinburgh, and saw two very bad operations, one on a child, but I rushed away before they were completed. Nor did I ever attend again, for hardly any inducement would have been strong enough to make me do so; this being long before the blessed days of chloroform.

Then, or earlier, Darwin developed a lifelong aversion to blood and a terror of illness of any kind. Unable to admit to his father his growing disinclination for medicine, he concentrated on what he did enjoy at medical school: natural history classes.

At the end of his first year, his older brother Erasmus left Edinburgh to continue his studies in London. The two had been so close that they had made scant efforts at forming outside friendships. Now, alone at school and farther from his family than he had ever been, Darwin was forced to look outward. It was good for him. He joined the Plinian Society, a club of like-minded undergraduates who met regularly to read and discuss papers on natural history. Through the society, he met its former secretary, Robert Grant, who had trained as a doctor but become a noted lecturer and respected naturalist. Reserved, austere, melancholic, a confirmed bachelor, and possibly a homosexual, Grant formed a succession of attachments with favorite students, often later falling out with them. For a time during his second year in Edinburgh, Darwin was one of these.

Grant, who lived in a house on the shore near Leith Harbour, introduced Darwin to marine zoology. Together they collected invertebrates—tiny, gelatinous, spongiform creatures—from rock pools and oyster shells and the muck of fishermen's hauls from the seabed. Grant's fascination for these organisms was contagious. He brought them alive for Darwin, showing him the nature and context of their microworlds: how they lived, adapted, and metamorphosed; how they reproduced; and how to dissect them in seawater under a microscope. In that second year at Edinburgh, the medium of the sea became for Darwin one great microscope slide—a lens that held up to view the macro struggle for existence in the alternate world beneath the waves—and for the rest of his life he remained fascinated by tiny sea creatures.

Grant also talked with Darwin about the heretical theory of evolution. "He one day, when we were walking together, burst forth in high admiration of Lamarck and his views on evolution," Darwin wrote much later. In 1800, Frenchman Jean-Baptiste

Lamarck suggested that Earth's species had not been created in their only and unalterable form at the biblical dawn of Creation, but that they had gradually and continually altered, adapting to a constantly changing environment, becoming and generating new species by transmutation. This was a direct contradiction of the Bible, and in an earlier age Lamarck would have been burned at a stake for his views. In Lamarck's time, most people, even forward-looking scientists, still believed that God had created Earth, and all life upon it, as a "Great Chain of Being" from the smallest—that is, lowest—creatures, to the highest, Man, with each species occupying its own predetermined, unchangeable link in that chain. Lamarck's claim that creatures evolved from lower to higher—and that Man had also evolved from lower forms, most recently apes—was a blasphemy. But such views were not new to Darwin. His own grandfather, the first Erasmus Darwin (1731–1802), a doctor and a naturalist, had been a famous evolutionary thinker in England before Lamarck and had expressed his ideas in the form of popular, if controversial, poetry:

> *Organic Life beneath the shoreless waves*
> *Was born and nurs'd in Ocean's pearly caves;*
> *First forms minute, unseen by spheric glass,*
> *Move on the mud,or pierce the watery mass;*
> *Then as successive generations bloom,*
> *New powers acquire and larger limbs assume.*

Views such as these, and Lamarck's, were heretical thinking in the early nineteenth century. Most scientists saw the clear hand of God in the design and perfection of a well-ordered Heaven and Earth, not a world of destructive change that wiped out whole species through a process of slow attrition. That would surely be a godless Universe, a living anathema, the third, hellish panel of a Bosch triptych. But a few did believe exactly this, and Robert Grant was one of them.

Darwin's studies with Grant produced his first scientific paper, read to the Plinian Society on March 27, 1827. He had observed through a poor microscope what apparently had not been seen or noted by anyone else: the frenzied swimming of tiny eggs that explained the fertilization of the species *Flustra*, a seaweed-like creature. Darwin was thrilled by what appeared to be a first, but his pleasure was shortlived. Grant appropriated his findings, without crediting Darwin's efforts, passing them off as his own observations in a paper read to the more august Wernerian Natural History Society on March 24, 1827, three days before Darwin's presentation to the Plinian Society. Years later, Darwin told his daughter Henrietta that his first scientific discovery had provided his first glimpse of "the jealousy of scientific men."

Relations between professor and student cooled, but Grant's influence on Darwin was profound: it was his first exposure to the deep inductive exploration of a science. The science became the foundation of Darwin's evolutionary theories, and the scientist provided him with a model for obsession.

At the end of the school year, Darwin came home and told his father he could not continue his studies to be a doctor. Dr. Darwin was furious. "You care for nothing but shooting, dogs, and rat-catching, and you will be a disgrace to yourself and all your family." Darwin later agreed with him: "He was very properly vehement against my turning into an idle sporting man, which then seemed my probable destination."

Darwin's problem was that he had no ambition. The Darwins were landed gentry, wealthy property owners, and there was no financial incentive for Charles to find a career; he would always be wealthy. He enjoyed shooting and hunting more than anything else. In between such outings, he liked collecting and studying small creatures. It was all he wanted to do. But Robert Darwin wasn't going to see his son turn into a wastrel and dilettante, so he told him to prepare for the church.

The profession of clergyman was just as respectable as being a medical man, requiring much the same sort of bedside manner.

The Church of England was then a gentleman's club with the most impeccable credentials, and in many places extremely well-appointed. Parish ministers were provided with houses, an adequate income (which Darwin could amply supplement), and instantly acquired social status. They had servants, bred fine dogs and horses, and maintained good wine cellars. The role had the same comfortable, tweedy informality as a schoolmaster's, but with a higher social profile and a lot more leisure time. For the vicar-with-a-hobby it was a platform upon which to develop a full-blown avocation. Many of the notable writers and scientists of the late eighteenth and early nineteenth century were clergymen who studied and wrote books during their great stretches of spare time. An appointment in the right place would be perfect for him, allowing him to shoot with the gentry, ride with the local hunt, botanize, geologize, collect to his heart's content, write papers and monographs, and achieve status in the scientific community, if that's where his enthusiasm led him. Such a man had been the eighteenth-century parson William Paley, whose book, *Natural Theology, or Evidences of the Existence and Attributes of the Deity* had become a standard textbook for theological students. In 1798, Thomas Malthus, another country parson, wrote *An Essay on the Principle of Population* that would become one of the most influential works of the next fifty years—it would play a crucial role in the evolution of Darwin's later thinking. Here were the perfect role models for a scientifically distracted clergyman.

Darwin was happy with his father's sensible suggestion, as long as he could carry it off in good conscience.

I asked for some time to consider, as from what little I had heard and thought on the subject I had scruples about declaring my belief in all the dogmas of the Church of England; though otherwise I liked the thought of being a country clergyman. Accordingly I read with care Pearson on the Creed and a few other books on divinity; and as I did not then in the least doubt the strict and literal truth of every word in the Bible, I soon persuaded myself that our Creed must be fully accepted.

The training would not be difficult. A bachelor of arts degree at a university, followed by a period of divinity study, would get him his holy orders. In January 1828, he went to Christ's College, Cambridge.

There he started having fun. He met a like-minded cousin, William Darwin Fox, also studying for holy orders, and the two were soon spending most of their time together, rambling through the countryside on collecting expeditions and doing only just enough work to pass their exams.

Another major preoccupation at Cambridge was to serve him as well as any of his studies. Darwin had become a crack shot at age fifteen, and he loved shooting a rifle more than anything else. "How I did enjoy shooting," he wrote later. "If there is bliss on earth, that is it."

My zeal was so great that I used to place my shooting boots open by my bed side when I went to bed, so as not to lose half a minute in putting them on in the morning. . . . When at Cambridge I used to practise throwing up my gun to my shoulder before a looking glass to see that I threw it up straight. Another and better plan was to get a friend to wave about a lighted candle, and then to fire at it with a cap on the nipple, and if the aim was accurate the little puff of air would blow out the candle.

Darwin studied books about guns and the practice of shooting. He kept a "game book," a ledger of everything he shot, and lists of what bores of shot were right for different game. He went on shooting parties with Fox and other Cambridge students. It was as much an accepted and desirable part of a young gentleman's training for life as anything else, and probably of more subsequent value to Darwin than any of his academic studies.

The only competition for the long hours and days spent shooting was an obsession he picked up from his cousin William Fox.

> No pursuit at Cambridge was followed with nearly so much eagerness or gave me so much pleasure as collecting beetles. It was the mere passion for collecting, for I did not dissect them and rarely compared their external characters with published descriptions, but got them named anyhow. . . . No poet ever felt more delight at seeing his first poem published than I did at seeing in Stephens' *Illustrations of British Insects* the magic words, "captured by C. Darwin, Esq."

Darwin was made for entomology. It sent him outdoors in all weathers, on foot or horseback, to spend hours with friends and dogs, kicking over fallen logs and feeling at the same time, at last, useful. It was early days in the natural sciences, and a dedicated amateur could soon gather a collection of mounted insects that could rival a museum's. Darwin's enthusiasm was all-consuming.

> One day on tearing off some old bark, I saw two rare beetles and seized one in each hand; then I saw a third and new kind, which I could not bear to lose, so that I popped the one that I held in my right hand into my mouth. Alas it ejected some intensely acrid fluid, which burnt my tongue so that I was forced to spit the beetle out, which was lost, as well as the third one.

Darwin began his beetle hunting at the beginning of the great Victorian mania for collecting—rocks, fossils, ferns, seashells, natural objects of every possible kind, to be taken home, cataloged, occasionally discovered and named, mounted and set up on boards and in cabinets for display. It was an era of newness in the natural sciences when only the degree of industry separated the enthusiastic amateur from the expert.

The rage for local discoveries produced a wonderful range of collecting jars, tins, nets, and, most importantly, clothing and accoutrements for purchase by the generally well-to-do classes that had the leisure time and the money to follow, and be seen to follow, such pursuits. Darwin naturally outfitted himself completely.

"He would have made you smile," wrote John Fowles of a young Victorian gentleman, another Charles, and the clothes he wore for a geologizing walk along a beach, in his novel *The French Lieutenant's Woman.*

> He wore stout nailed boots and canvas gaiters that rose to encase Norfolk breeches of heavy flannel. There was a tight and absurdly long coat to match; a canvas wideawake hat of an indeterminate beige; a massive ashplant, which he had bought on his way to the Cobb; and a voluminous rucksack, from which you might have shaken out an already heavy array of hammers, wrappings, notebooks, pillboxes, adzes and heaven knows what else. Nothing is more incomprehensible to us than the methodicality of the Victorians; one sees it best (at its most ludicrous) in the advice so liberally handed out to travelers in the early editions of Baedecker. Where, one wonders, can any pleasure have been left? How, in the case of Charles, can he not have seen that light clothes would have been more comfortable? That a hat was not necessary? That stout nailed boots on a boulder-strewn beach are as suitable as ice skates?
>
> Well, we laugh. But . . . if we take this obsession with dressing the part, with being prepared for every eventuality, as mere stupidity, blindness to the empirical, we make, I think, a grave— or rather a frivolous—mistake about our ancestors; because it was men not unlike Charles, and as overdressed and overequipped as he was that day, who laid the foundations of all our modern science. Their folly in that direction was no more than a symptom of their seriousness in a much more important one. They sensed that current accounts of the world were inadequate; that they had allowed their windows on reality to become smeared by convention, religion, social stagnation; they

knew, in short, that they had things to discover, and that the discovery was of the utmost importance to the future of man.

His disinterest in more formal studies notwithstanding, these pursuits brought Darwin into close contact with other naturalists, notably professors Henslow and Sedgwick. He became a part of their coteries, joined them for field trips, attended Henslow's soirées. He began to feel his true calling, whether it was something that could be properly considered a profession or not. He began, with considerable excitement, to think of himself as a scientist. He read of naturalists who had ranged far beyond the fen country around Cambridge: von Humboldt's account of his adventures in Tenerife and Brazil inflamed his imagination. He longed to travel.

Darwin got the news of the voyage around the world after walk-ing across north Wales for three weeks on a geological tour with Cambridge Professor Adam Sedgwick, who was hoping to correct and add to George Greenough's 1820 map of the geology of England and Wales. He had invited young Darwin along to help him, and also to give Darwin a chance at some practical field geologizing before the Tenerife expedition that he liked to talk about.

The two left from Darwin's family home in Shrewsbury on August 5th, riding in Sedgwick's carriage to Llangollen in north Wales. From there they walked along the bald rock-rimmed Vale of Clwyd toward Caernarvon on the coast. At Saint Asaph, Sedgwick sent Darwin off on his own to look for signs of a stratum of Old Red Sandstone shown on Greenough's map. When they met up again that evening in Colwyn, neither had seen a trace of Old Red. Sedgwick told Darwin that the structure of the Vale of Clwyd would now be revised on the basis of their work. Impressed and grateful for Sedgwick's trust in him, Darwin was "exceedingly proud."

He returned to his family home, The Mount, on August 29, to find a note from Henslow accompanying the now dog-eared letter from George Peacock. The invitation was breathtaking; it laid before Darwin opportunities that his training in natural science was only beginning to enable him to imagine. At a stroke it eclipsed von Humboldt's Tenerife and Brazil. As a peruser of traveling books, he had probably also seen an edition of Captain James Cook's accounts of his circumnavigations, with drawings and watercolors by his expedition artists William Hodges and John Webber. These (which may be seen today in the National Maritime Museum in Greenwich, England) portray the Englishmen's idealized views of the noble savages amid the sylvan sublimities of their natural settings in Polynesia and the remote, majestic fjordland of New Zealand—Paradise about to be lost—at the moment of contact with Europeans. Here Man and Nature were believed to still exist in the Edenic state, and fifty years after Cook the untrammeled world was thought to be—and surely was—a naturalist's paradise.

As soon as he read the letters, Darwin told his three sisters he would go. But his father was so strongly against the idea that Darwin almost immediately gave it up. The next morning he wrote to Peacock and Henslow, thanking them but regretfully turning down the offer.

> As far as my own mind is concerned, I should think, <u>certainly</u>, most gladly have accepted the opportunity, which you have so kindly offered me, [he wrote to Henslow] — But my Father, although he does not decidedly refuse me, gives such strong advice against going, — that I should not be comfortable, if I did not follow it. — My Fathers objections are these; the unfitting me to settle down as a clergyman, — my little habit of seafaring, — the <u>shortness of the time</u> & the chance of my not suiting Captain FitzRoy. — It certainly is a very serious objection, the very short time for all my preparations, as not only body but mind wants making up for such an undertaking. — But if it had not been for my Father, I would have taken all risks. . . . Even if I

was to go, my Father disliking would take away all energy, & I should want a good stock of that. — Again I must thank you; it adds a little to the heavy, but pleasant load of gratitude which I owe to you. —

Darwin lived securely and happily wrapped in the community of his family, the second youngest of six children, cosseted and fussed over by his three older sisters who formed the maternal bulwark in his life after the early death of his mother when he was eight. His father was a benign but magisterial presence at the head of the household, a physically immense man, rich, intelligent, influential, a creator of the world around him, a figure of absolute authority to his son. Charles was not a headstrong young man who would go against such a father's wishes. When Dr. Darwin expressed his disapproval, Charles didn't argue.

Instead, he packed a gun and rode to his Uncle Jos's house. Darwin and Josiah Wedgwood II enjoyed a relationship almost as close as father and son. Uncle Jos was not as formidable a man as Robert Darwin, not the authoritarian in Charles's life, yet he was a strong influence on his nephew. The two enjoyed shooting together so much that Darwin called Maer, the Wedgwood home, "Bliss Castle." Dr. Darwin respected Wedgwood, his brother-in-law, and the influence he had over his son. As Darwin was leaving for Maer, his father told him pointedly: "If you can find any man of common sense who advises you to go, I will give my consent." And he gave Darwin a note to take to his uncle:

Charles will tell you of the offer he has had made to him of going for a voyage of discovery for 2 years. — I strongly object to it on various grounds, but I will not detail my reasons that he may have your unbiassed opinion on the subject, & if you think differently from me I shall wish him to follow your advice.

Darwin and his Uncle Jos talked it over. He put his father's objections on paper for Wedgwood to look at and mull over.

1) Disreputable to my character as a Clergyman hereafter
2) A wild scheme
3) That they must have offered to many others before me, the place of Naturalist
4) And from its not being accepted there must be some serious objection to the vessel or expedition
5) That I should never settle down to a steady life hereafter
6) That my accomodations would be most uncomfortable
7) That you should consider it again as changing my profession
8) That it would be a useless undertaking

Wedgwood—the "man of common sense" whom Dr. Darwin was plainly referring to—wrote him a letter addressing each of his eight objections:

My dear Doctor

I feel the responsibility of your application to me on the offer that has been made to Charles as being weighty, but as you have desired Charles to consult me I cannot refuse to give the result of such consideration as I have been able to give it. Charles has put down what he conceives to be your principle objections & I think the best course I can take will be to state what occurs to me upon each of them.

1—I should not think it would be in any degree disreputable to his character as a clergyman. I should on the contrary think the offer honorable to him, and the pursuit of Natural History, though certainly not professional, is very suitable to a Clergyman.

2—I hardly know how to meet this objection, but he would have definite objects upon which to employ himself and might acquire and strengthen, habits of application, and I should think would be as likely to do so in any way in which he is likely to pass the next two years at home.

3—The notion did not occur to me in reading the letters & on reading them again with that object in mind I see no ground for it.

4—I cannot conceive that the Admiralty would send out a bad vessel on such a service. As to objections to the expedition, they will differ in each mans case & nothing would, I think, be inferred in Charles's case if it were known that others had objected.

5—You are a much better judge of Charles's character than I can be. If, on comparing this mode of spending the next two years, with the way in which he will probably spend them if he does not accept this offer, you think him more likely to be rendered unsteady & unable to settle, it is undoubtedly a weighty objection—Is it not the case that sailors are prone to settle in domestic and quiet habits.

6—I can form no opinion on this further than that, if appointed by the Admiralty, he will have a claim to be as well accomodated as the vessel will allow.

7—If I saw Charles now absorbed in professional studies I should probably think it would not be advisable to interrupt them, but this is not, and I think will not be, the case with him. His present pursuit of knowledge is in the same track as he would have to follow in the expedition.

8—The undertaking would be useless as regards his profession, but looking upon him as a man of enlarged curiosity, it affords him such an opportunity of seeing men and things as happens to few.

You will bear in mind that I have had very little time for consideration & that you and Charles are the persons who must decide.

Wedgwood sent the letter off early on September 1. He and Darwin then tried to distract themselves shooting partridge—it was the first day of the season, an opportunity neither could ignore—but their hearts weren't in it and they made a poor bag. At 10 A.M., Wedgwood impatiently decided he wanted to talk to the doctor face to face, and uncle and nephew set out in his carriage for The Mount, determined to present their opinion with persuasive arguments.

But they found Dr. Darwin in a compliant mood. He'd had a change of heart, perhaps reflecting on the honor of the invitation by the Admiralty, passed to his son by illustrious professors who clearly thought highly of him. Perhaps his greatest objection was one not mentioned in Charles's list: he feared for his son's life on a voyage to the wilder parts of the little known world. Captain Cook himself, despite the protection of a boatload of armed seamen, had been killed in the Sandwich Islands (Hawaii) by natives who had previously revered him. In the end Dr. Darwin recognized the singular opportunity that was presenting itself, and he had decided to let Charles go—with all the considerable financial assistance and encouragement he could provide.

But by then, it appeared, FitzRoy had asked someone else.

II

Or so FitzRoy told Alexander Charles Wood, his cousin, a Cambridge undergraduate and one of Peacock's students. Wood had heard about the voyage and the opportunity for a naturalist and had written to FitzRoy enthusing about Darwin, but pointing out that he was a Whig (a liberal). FitzRoy wrote back informing Wood that the position had already been filled. Wood told Henslow. Darwin, who had by then returned to Cambridge to consult with Henslow about the coming voyage, was dismayed. Both he and Henslow were now confused, and Henslow angrily felt that Peacock had misrepresented the availability of the position, getting everybody worked up over a tantalizing maybe.

FitzRoy may indeed have asked a friend, a "Mr. Chester," as he later told Darwin, or he may very reasonably have been hedging his bets until he met the candidate produced by the long arm of the Admiralty and Cambridge. As much as he felt the need for a gentleman companion on such a long voyage, he knew the peril of taking the wrong gentleman, of being locked in close quarters for months and years on end with the wrong personality. Nevertheless, Darwin and FitzRoy arranged to meet in London and they kept the date.

They were both young men—FitzRoy 26, Darwin 22—with

the world and a ship at their disposal. They met and appraised each other as potential partners in a very great adventure. But FitzRoy had the experience, accomplishment, and self-confidence of someone much older, and he was the captain—a most absolute authority.

FitzRoy stressed to Darwin the rigors of the voyage: the danger, the storms, the extremes of cold and heat, the risk of illness, the unvarying diet—he probably described these at their grimmest. Also, the voyage might not result in a circumnavigation, he told Darwin. His first duty was the completion of the South American survey, and the time taken for this might prevent them continuing on around the world. He painted as unenticing a picture as possible.

As they talked, they sized each other up. Both men were on their best behavior, and after they had spoken a while and relaxed, both liked what they saw in the other. Like any sailor, FitzRoy would certainly have told Darwin stories of his earlier voyage in the *Beagle*, of Cape Horn and its weather, of the trials of surveying under such conditions. Undoubtedly he told him about the Fuegians, and the three "improved" specimens he was returning to Tierra del Fuego. All of this, the raw material of adventure, could only have whetted Darwin's appetite. He was enormously impressed by FitzRoy's manner, his directness, his authority, his intelligence, and probably above all his grasp of science, in which area the captain was then far more knowledgeable than the young Cambridge graduate.

FitzRoy in turn was charmed by Darwin, by his enthusiasm and his own well-developed knowledge. More importantly, he perceived in Darwin the breed of companion he sought. Here was a young man who knew horses and guns, who had dined all his life in the company of thinkers and gentlemen, who was expansive of mind and could probably be counted on as an affable and intelligent dinner companion for a thousand and more nights at the same table. FitzRoy also knew of Erasmus Darwin, the young man's grandfather, the poet and evolutionary thinker.

And he had come with the recommendations of professors and admirals.

The only thing that bothered FitzRoy was Darwin's face. His beliefs in the telling aspects of the shape of the cranium and facial features applied to Englishmen just as valuably as to savages, and Darwin's hooded brow and large, spatulate nose gave FitzRoy serious pause. "He doubted whether anyone with my nose could possess sufficient energy and determination for the voyage," Darwin later wrote.

But this first meeting was encouraging to both. They agreed to dine together that same evening. That afternoon, Darwin wrote to his sisters.

> I have seen him; it is no use attempting to praise him as much as I feel inclined to do for you would not believe me. One thing I am certain, nothing could be more open and kind than he was to me. . . . He says nothing would be so miserable for him as having me with him if I was uncomfortable, as in a small vessel we must be thrown together, and thought it his duty to state everything in the worst point of view: I think I shall go on Sunday to Plymouth to see the vessel. There is something most extremely attractive in his manners and way of coming straight to the point. If I live with him, he says, I must live poorly—no wine, and the plainest of dinners. . . . I like his manner of proceeding. He asked me at once, "Shall you bear being told that I want the cabin to myself? when I want to be alone. If we treat each other this way, I hope we shall suit; if not probably we should wish each other at the Devil." . . . I am writing in a great hurry . . . I dine with him today.

The dinner was a success. Darwin wrote again to his sisters the next day.

> I write as if . . . it was settled, but it is not more than it was, excepting that from Capt. FitzRoy wishing me so much to go, and from his kindness. I feel a predestination I shall start. I

spent a very pleasant evening with him yesterday . . . he is of a slight figure, and a dark but handsome edition of Mr Kynaston, and, according to my notions, pre-eminently good manners. . . . This is the first really cheerful day I have spent since I received the letter, and it is all owing to the sort of involuntary confidence I place in my beau ideal of a Captain.

The following weekend the two young men traveled by steam packet from London to Plymouth, to see the *Beagle* in the Devonport dockyard. Darwin could only have gained the poorest grasp of what she would be like to live aboard: the ship was without masts or interior bulkheads, and he thought she looked like a wreck. His own quarters in the chartroom aft were cluttered with carpenters and appeared dismayingly small, but FitzRoy assured him that he would make him comfortable and provide him with a workshop. The voyage around the coast had been pleasant, and Darwin's initial worries that he might suffer from seasickness were calmed.

It was settled. FitzRoy wrote to Beaufort that he approved of Darwin and wanted him aboard as naturalist. He told Darwin to do and buy what he must for the voyage and to report on board the *Beagle* by the end of October; he had chosen November 4 as their tentative departure date.

Then he set about preparing the ship.

After six years of service, most of the *Beagle*'s planking and deck frames were rotten and had to be replaced. As this work, along with a general refit of all her gear commenced at Devonport, FitzRoy got permission to make some other changes. When he had filled the "death vacancy" made by Captain Stokes and boarded the *Beagle* in Rio de Janeiro in 1828, FitzRoy had done his best with what he was given. Now, after two seasons in the far south and 20,000 miles of voyaging, he had definite ideas for the vessel's improvement. He had the upper deck raised eight inches throughout the ship, giving increased headroom for the whole crew, who had formerly moved about belowdecks in an uncomfortable stoop. This not only made life below more comfortable for everyone (a condition FitzRoy placed great value upon), but it added to the ship's buoyancy, literally providing more air inside the hull to resist a capsize when pushed over onto her beam by wind and seas. (Raising the deck would also have lifted the vessel's center of gravity, possibly increasing her tendency to roll, making her less stable, but nobody seems to have noticed or remarked on this.) FitzRoy was pleased with the alteration, which "proved to be of the greatest advantage to her as a sea-boat, besides adding so materially to the comfort of all on board," he later wrote.

Heading out across a world of storms, FitzRoy—a tireless follower of modern scientific developments—installed lightning conductors of a type recently invented by William Snow Harris, a Fellow of the Royal Society. These were heavy-gauge continuous copper strips built into the masts and spars and bowsprit, running from the very top of the ship down through the rigging, down the outside of the hull into the ship's copper-sheathed bottom for grounding in the water. Struck by lightning a number of times in the voyage to come, the *Beagle* never received any damage.

Other improvements included a new type of rudder, designed by Captain Lihou of the Royal Navy, which enabled the rudder's pintle, or hinge, to be replaced if broken on a distant shore. The galley's open fireplace was torn out and "one of Frazer's stoves" with an oven installed in its place. This was more efficient in its consumption of fuel and made cooking on board safer in all weathers.

While these intrinsic shippy improvements were being carried out, FitzRoy turned to the equipping of his vessel for scientific purposes. Both he and Beaufort, with the full weight of the Admiralty behind them, were determined to provide the *Beagle* with the best, most up-to-date equipment and instruments possible.

Preeminent among these, as necessary as a compass, were the ship's chronometers. Accurate time was vital to accurate navigation, the determination of longitude, and the consequential correct placing of rocks, islands, and coasts in FitzRoy's surveys—his primary mission. Since the groundbreaking work of John Harrison a century earlier, which had won him the government's £20,000 award for a chronometer that would provide mariners with longitude, timepieces had been continually refined. A chronometer is not a clock that keeps perfect time, but one whose mechanism gains or loses it at a nearly consistent, measurable rate. Thus once set with the correct time at Greenwich, that Greenwich Mean Time—the basis for all celestial navigation—can still be accurately known weeks later, far out at sea, after the correction for loss or gain is applied. That was the theory. But a chronometer, like any mechanical device, can go wrong or break down, so several were carried aboard ships, their rates of loss or gain observed and compared with each other, and an average assumed.

Twenty-two chronometers were carried aboard the *Beagle* on her second voyage. The Admiralty provided sixteen, and FitzRoy, in his pursuit of absolute sufficiency, even redundancy, bought a further six. They were the finest money could buy, each a jewel of mechanical contrivance, purchased from different clockmakers. Their storage aboard the ship and the conditions of their use were as important as their construction for keeping good time. Each was housed in a small wooden box, suspended by well-oiled gimbals that kept the clock level despite the rocking of the ship and its own box around it. The twenty-two boxes were kept in a small cabin beside FitzRoy's own, placed in sawdust three inches thick beneath and at the sides of each box as a shock absorber.

> Placed in this manner [wrote FitzRoy], neither the running of men upon deck, nor firing guns, nor the running-out of chain cables, caused the slightest vibration in the chronometers, as I often proved by scattering powder upon their glasses and watching it with a magnifying glass, while the vessel herself was vibrating to some jar or shock.

A good fix on this punctilious scientist-captain is provided by imagining him in a gale at sea, oiled coat thrown off, head toweled to avoid dripping, bent over his chronometers as the ship pitches around him, magnifying glass in hand, to observe the fine sprinkling of powder on the glass tops in his rookery of clocks.

The greatest disturbance to the clocks was their daily winding. To tend them, and ensure a consistency of attention and handling, FitzRoy hired a supernumerary to the ship's crew, George James Stebbing, eldest son of a mathematical instrument maker from Portsmouth, who came along for the entire voyage. Only Stebbing touched the clocks, winding them every morning at 9 A.M. He came again at noon daily to compare their times and rate their gains and losses. Only FitzRoy and Stebbing ever entered the small cabin of chronometers. Most of the boxes were never moved after first being secured in sawdust in 1831 until the completion of the voyage in 1836.

Between voyages, FitzRoy had his sextant, theodolite, small portable compasses, telescopes of several powers, and other instruments cleaned and repaired as necessary. New ones were purchased for this voyage. Once aboard ship, Stebbing looked after these too.

As with the extra chronometers, FitzRoy paid for Stebbing out of his own pocket.

The whaleboat built by Jonathan May on Chiloé Island in 1829, stolen by Fuegians a few months later, is one of the great lost artifacts of history. FitzRoy couldn't have recognized it as such, but the first voyage had nonetheless given him the keenest appreciation for a sufficient quantity of ship's boats. It may have occurred to him that this second voyage of the *Beagle* might never have happened if he'd had more boats on the first. Six new boats were now built: two 25-foot whaleboats, two 28-foot whaleboats, a 23-foot cutter, and a 26-foot yawl. Their tough hulls were made from a double layer of diagonal planks, a method refined by a foreman at

Plymouth Dockyard, a Mr. Jones. The Admiralty felt four were enough, but FitzRoy disagreed. He paid for the other two. These were carried on deck, the cutter nesting inside the yawl in the forward waist, the two larger whaleboats carried upside down aft, the smaller whaleboats held outboard on davits aft. A seventh boat, a 15-foot dinghy (*dingy* is an Indian word for small boat) was carried in davits astern. These boats greatly interfered with movement on deck, and the handling of the rig, but they were an accepted part of any ship's gear and the sailors scrambled over and around them.

Although Frenchman Louis Daguerre was already experimenting with capturing permanent images on silver-plated sheets of copper used with a camera obscura, photography was still virtually unknown. To make a visual record of the voyage, FitzRoy hired another civilian supernumerary, an artist, Augustus Earle, and, as with Stebbing, paid for him out of his own pocket. The son of an American painter living in England, Earle had studied at the Royal Academy in London. But he was unable to settle into the role of society painter; afflicted by a serious wanderlust, he had already traveled to Australia, New Zealand, India, and South America as an itinerant artist, earning a scratchy living painting portraits of colonial governors. When FitzRoy met him in London he was thirty-seven years old, peddling his paintings of Maoris and finding no takers. His lack of success was probably because Earle had an eye for the truth in a subject, rather than for what might be pleasing to European taste. His subsequent drawings and paintings of the *Beagle* and the people and places it encountered are among the most valuable and accurate records of nineteenth-century exploration. FitzRoy could not have chosen better.

(Two years later, Earle's health forced him to leave the *Beagle* in South America. FitzRoy replaced him with Conrad Martens, a landscape painter. "By my faith in bumpology, I am sure you will like him," FitzRoy wrote to Darwin, who was away from the ship, traveling across Argentina at the time. "His landscapes are *really* good." FitzRoy's phrenological assessment of the new artist was justified. Martens's watercolors of the *Beagle* are now famous.)

The value of certain foods in combatting scurvy, the age-old scourge of sailors cut off from fresh supplies and restricted to the traditional and revolting seagoing diet of weevil-filled ship's biscuit, rotten pork, and salt beef, was by then well-known. FitzRoy loaded his ship with the healthiest foods that could be carried in bulk, by the latest methods of preservation, for long periods in the airless confines of a ship's hull. "Among our provisions were various antiscorbutics—such as pickles, dried apples, and lemon juice—of the best quality, and in as great abundance as we could stow away; we also had on board a very large quantity [5000–6000 cans] of Kilner and Moorsom's preserved meat, vegetables, and soup; and from the Medical Department we received an ample supply of antiseptics."

Nothing was spared the *Beagle* and the preparations for her voyage. The Admiralty and the naval dockyard workers at Devonport gave her their best efforts and materials, and FitzRoy saw that she was as completely equipped for scientific inquiry as the age allowed. The *Beagle* sailed as loaded with the cutting-edge technology of her time as any rocket ship that ever blasted off into space. She was an ark of discovery on her own five-year mission.

"Perhaps no vessel ever quitted her own country with a better or more ample supply . . . of every kind of useful provision," wrote FitzRoy. He paid for much of it. His inheritance that came from his family was sufficient to sustain a comfortable gentleman's existence ashore, but FitzRoy had better ideas for its use. The expenses he personally undertook for the voyage were considerable, and to pay for them he dug deep into his capital. He did this in the belief that the excellence of results he hoped to obtain would in time reward and reimburse him. It was a conscientious investment in his career and future, and in 1831 it looked a good one. Robert FitzRoy was in so many ways the finest product of his times. Blessed with wealth and station, he didn't waste his life, as did so many of his peers, hunting, shooting, whoring, and gambling. He joined the navy and made the utmost of the opportunity it afforded him to add not only to his own learning, but to the pool of knowledge that was then beginning to flood the scientific community.

FitzRoy and the *Beagle* were made for each other. Together they provided Charles Darwin with a unique springboard for his own leap to destiny.

12

Darwin traveled to Devonport by coach on October 24, 1831. "Arrived here in the evening after a pleasant drive from London," was his first entry in the diary he had decided to keep during the voyage. He put up at a hotel and went aboard the *Beagle* the next day.

> 25th   Went on board the Beagle, found her moored to the Active hulk & in a state of bustle & confusion. — The men were chiefly employed in painting the fore part & fitting up the Cabins. — The last time I saw her was on the 12th of Septr she was in the Dock yard & without her masts or bulkheads & looked more like a wreck than a vessel commissioned to go round the world.

> 26th   Wet cold day, went on board, found the Carpenters busy fitting up the drawers in the Poop Cabin. My own private corner looks so small that I cannot help fearing that many of my things must be left behind.

"Went on board" was his only entry for November 2. Darwin was also writing daily letters to his family and friends, and his better efforts at description went into these. The diary remained, for a time, an uninspired itemization of the main points of his day. His style, expression, and the attention he gave to his daily log were to improve, however: on publication in 1839, his *Journal of Researches*, fashioned from this diary, became an instant bestseller, and has remained so for 165 years.

The *Beagle* had been due to sail in early November, but preparations delayed her departure for more than a month, and then westerly gales in the English Channel kept her pinned ashore. At first Darwin amused himself observing the frantic activity aboard the ship, and walking to and from Plymouth with FitzRoy and some of the junior officers, sometimes helping them with their abstruse preparations.

> Monday 31st (October)   Went with Mr Stokes [mate and assistant surveyor John Lort Stokes, who had been a midshipman on the *Beagle*'s first voyage] to Plymouth and staid with him whilst he prepared the astronomical house belonging to the Beagle for observations on the dipping needle. . . .
>
> 4th (November)   Cap FitzRoy took me in the Commissioners boat to the breakwater, where we staid for more than an hour. Cap. FitzRoy was employed in taking angles, so as to connect a particular stone, from which Cap King commenced for the last voyage his longitudes, to the quay at Clarence Baths, where the true time is now taken.

Here, FitzRoy was doing for geography what his chronometers did for time, and what the zero meridian of longitude at Greenwich does for east and west around the world: he was starting with a known position, or value, on which to base all subsequent measurements. His chronometer and sextant readings for the next five years, all his determinations of longitude, and positions of rocks, headlands, shorelines, islands, and continents, would be laid down according to the geographic relationship they bore to a stone in Plymouth's breakwater. That arbitrary absolute provided the key to a graspable shape of the whole world.

As busy as he was, FitzRoy did not neglect one aspect of his private life: in Plymouth he met a young woman he liked, Mary O'Brien, the daughter of a local gentleman and army major general. She liked him too. The dashing young captain visited the O'Brien house as often as he could, but he kept it to himself; not even Darwin was aware of his interest in anything beyond the demands of the coming voyage.

Darwin continued trying to make himself useful, or at least amuse himself, but this became more difficult as the weeks in port wore on.

5th (November)   Wretched, miserable day, remained reading in the house.

6th   Went with Musters to the Chapel in the Dock-yard. — It rained torrents all the evening. . . .

Monday 7th   Staid at home.

8th   In the morning marked time [i.e., Darwin noted the time] whilst Stokes took the altitude of the sun. — Went on board the Beagle; she now begins for the first time to look clean & well arranged.

10th   Assisted Cap. FitzRoy at the Athaeneum in reading the various angles of the dipping needle. . . .

11th   Went again to the Athaeneum & spent the whole day at the dipping needle. — The end, which it is attempted to obtain, is a knowledge of the exact point in the globe to which the needle points. This means of obtaining it is to take, under all different circumstances, a great number of observations, & from them to find out the mean point. — The operation is a very long & delicate one. . . .

12th   Took a walk to some very large Limestone quarries, returned home & then went on board the Beagle. — The men had just finished painting her. . . . For the first time I felt a fine naval fervour; nobody could look at her without admiration.

13th   Walked to Saltram & rode with Lord Borrington to
Exmoor to see the Granite formation. . . .

15th   Went with Cap FitzRoy to Plymouth & were
unpleasantly employed in finding out the inaccuracies of
Gambeys new dipping needle.

17th   A very quiet day.

18th   Cap FitzRoy has been busy for these last two days with
the Lords of the Admiralty.

20th   Went to Church & heard a very stupid sermon, & afterwards took a long walk. . . .

Darwin was more than simply bored. He was suffering crushing
loneliness and apprehension. With too much time on his hands he
was becoming anxious about the coming voyage. He worried
about what might happen to him while he was away, and to loved
ones at home. He developed a rash—probably cold sores—around
his mouth. Always a vivid hypochondriac, he now became worried
about his heart and began to believe he would be too sick to sail.
Years later he was able to see this in perspective.

These two months at Plymouth were the most miserable which I
ever spent, though I exerted myself in various ways. I was out of
spirits at the thought of leaving all my family and friends for so
long a time, and the weather seemed to me inexpressibly
gloomy. I was also troubled with palpitations and pain about
the heart, and like many a young ignorant man, especially one
with a smattering of medical knowledge, was convinced that I
had heart disease. I did not consult any doctor, as I fully
expected to hear the verdict that I was not fit for the voyage,
and I was resolved to go at all hazards.

Other passengers gathered at Plymouth. On November 13,
the Fuegians arrived by steam packet from London, accompa-

nied by their schoolmaster, Mr. Jenkins. They remained ashore until the *Beagle* sailed. There is little record of what they did in the weeks while they waited for the ship to sail. There may have been a few occasions when FitzRoy took them to visit curious local notables, admirals, and friends, but such visits would not have been on the scale of their former socializing. FitzRoy was fully occupied with his preparations for the coming voyage, and he was no longer as comfortable showing off his Fuegians as he had been. He was still keeping Fuegia Basket separated from her two male companions; she stayed in Weakley's Hotel in Devonport, while the other two were put up elsewhere. His primary concern with them now was to hustle them out of the country as quickly as possible.

With them from London came a young missionary, Richard Matthews, recruited by Reverend William Wilson of Walthamstow, who was to sail to Tierra del Fuego and help the newly civilized Fuegians establish a mission. Still a teenager, Matthews was afire with religious fervor and the opportunity to put himself to God's use. His older brother was a missionary in New Zealand, a compelling role model for the younger man. But New Zealand was a sylvan Eden being farmed and settled by English families, and its native Maoris were a fierce, intelligent, highly cultured people. No advice or descriptions of the missionary life from his brother, nor his acquaintance with the roughly Anglicized York, Jemmy, and Fuegia from Walthamstow, could have prepared Matthews for the reality of an existence by himself at the far storm-wracked edge of the world with only unreconstructed Fuegians for neighbors. But he was ready, he believed, quivering with zeal, and, thanks to the generosity of Christian well-wishers, fully equipped. Reverend Wilson had raised a subscription to supply Matthews and the Walthamstow Fuegians with the necessities to recreate a little piece of God-fearing England in that wild foreign place. Stowed aboard the *Beagle* in October were their supplies: chamber pots, tea trays, complete sets of crockery, soup tureens, beaver hats, and white linen, in addition to books, tools, and blankets. Space was found for all these items inside the *Beagle*'s crammed hold, with "some very fair jokes enjoyed by the seamen" who packed them aboard.

On Monday November 21, Darwin brought all his books and instruments aboard the ship. His quarters were the poop cabin, which he shared with Stokes, and midshipman Phillip Gidley King, the seventeen-year-old son of Captain King, who had commanded the *Adventure*. The tiny 10-foot by 10-foot cabin that these three shared and filled with all their belongings was almost entirely taken up by a large chart table and rows of drawers the carpenters had built into the fore and aft bulkheads and along the sides of the hull. Stokes slept in a bunk just outside the chart room, and King's and Darwin's only beds were hammocks slung over the chart table. The space in the cabin was tight and exacting of movement and behavior. Darwin, who was over 6 feet tall and had never before been restricted to such close quarters, was, in the beginning, dismayed by the lack of space.

22 November   Went on board & returned in a panic on the old subject want of room. returned to the vessel with Cap FitzRoy, who is such an effectual & goodnatured contriver that the very drawers enlarge on his appearance & all difficulties smooth away.

The rest of the ship was to be equally crammed full of people and their belongings. Seventy-three men and one girl—Fuegia Basket—were now aboard the *Beagle*, or gathered in Plymouth ready to join the ship. FitzRoy listed them as follows:

Robert FitzRoy	Commander and Surveyor
John Clements Wickham	Lieutenant
Bartholomew James Sulivan	Lieutenant
Edward Main Chaffers	Master
Robert MacCormick	Surgeon

George Rowlett	Purser
Alexander Derbishire	Mate
Peter Benson Stewart	Mate
John Lort Stokes	Mate and Assistant Surveyor
Benjamin Bynoe	Assistant Surgeon
Arthur Mellersh	Midshipman
Philip Gidley King	Midshipman
Alexander Burns Usborne	Master's Assistant
Charles Musters	Volunteer 1st Class
Jonathan May	Carpenter
Edward Hellyer	Clerk

Acting Boatswain; sergeant of marines and seven privates; thirty-four seamen and six boys.

Supernumeraries:

Charles Darwin	Naturalist
Augustus Earle	Draughtsman
George James Stebbing	Instrument Maker

Richard Matthews and three Fuegians; my own steward; and Mr Darwin's servant.

Darwin did not bring aboard a servant—that would have made seventy-four men—but at first used one of the ship's boys, Henry Fuller, to help him organize and prepare specimens collected ashore. Later, he employed seventeen-year-old Syms Covington as his full-time assistant, co-opting him from the ship's crew and paying him £60 per year. Covington became Darwin's all-purpose assistant, secretary, shooter and co-collector throughout the voyage, and remained in his employment ashore in England until 1839. He was rarely mentioned by Darwin, as servants dressing and feeding their employers and secretaries mailing their letters seldom are, but Covington made significant contributions to Darwin's work during and after the voyage. He is one in a long line of friends and employees whose work and interest became the foundation for Darwin's studies and eventual reputation.

•   •   •

On November 23, filled to her newly raised decks with stores and food—"not one inch of room is lost, the hold would contain scarcely another bag of bread," wrote Darwin—the *Beagle* left the dock at Devonport and sailed a mile to Barnet Pool near the entrance to Plymouth harbor. There she anchored to await sailing weather, preferably northeasterly winds to blow her down the Channel and out into the Atlantic. But this was weeks in coming. November passed, and through most of December southwesterly gales blew up the Channel, making departure impossible. The heavily laden *Beagle* rolled and pitched in her anchorage, giving Darwin concerns about seasickness. At first, he felt well.

> December 4th . . . In the morning the ship rolled a good deal, but I did not feel uncomfortable; this gives me great hope of escaping seasickness.

But his hopes were soon dashed.

> Monday 5th It was a tolerably clear morning & sights were obtained, so we are now ready for our long delayed moment of starting. — it has however blown a heavy gale from the South ever since midday, & perhaps we shall not be able to leave the Harbour. The vessel had a good deal of motion & I was as nearly as possible made sick. . . .

Sailing was postponed. Darwin went ashore to dine with his brother Erasmus, who had come down to Plymouth to see him off. He spent the night at an inn rather than return to the ship heaving at anchor. He went aboard the next morning, prepared to leave, but again the weather prevented sailing, and Darwin, overcome by seasickness, returned ashore. For the next few days, as the *Beagle* pitched and rolled in her weatherbound anchorage, Darwin alternately worked at arranging his gear in the poop cabin, and fled ashore when queasiness compelled him. He and

Erasmus took walks on Mount Edgcombe, overlooking Plymouth, talked and ate together, assuaging Darwin's apprehensions and dread of seasickness.

On December 10 the weather appeared settled at last and the *Beagle* weighed anchors and sailed at 10 A.M. But as soon as the ship was past Plymouth's breakwater, Darwin became sick and took to his hammock. In the evening a strong gale began to blow from the southwest—directly on the ship's nose—forcing it to labor to windward. In a modern yacht, beating down the English Channel toward the open sea, such conditions are vile: the boat will lurch and pound against every oncoming wave like a four-wheel-drive vehicle bouncing over sand dunes. The *Beagle*, heavily laden and unable to point close to the wind, would have lifted, plunged, and rolled sickeningly in the short steep seas whipped up over the shallow waters of the Channel. It was a cruel baptism for Darwin. "I suffered most dreadfully," he wrote, "such a night I never passed, on every side nothing but misery."

The *Beagle* could make no headway against such wind and weather. Rather than exhausting and demoralizing his crew to no purpose so early, FitzRoy turned the ship about and she rolled back downwind to Plymouth, anchoring again in Barnet Pool. Darwin immediately fled ashore with seaman Musters, "a fellow companion in misery," for a long walk.

More weeks passed, with the *Beagle* bottled up in her anchorage. FitzRoy could only bide his time in frustration, watching the skies and the weather. Darwin, until now so admiring of his captain in his journal and letters home, got a first look at his capricious temper. Accompanying FitzRoy in Plymouth one day, Darwin reported that

> . . . he was extremely angry with a dealer in crockery who refused to exchange some article purchased in his shop: the Captain asked the man the price of a very expensive set of china and said "I should have purchased this if you had not been so

disobliging." As I knew that the cabin was amply stocked with crockery, I doubted whether he had any such intention; and I must have shown my doubts in my face, for I said not a word. After leaving the shop he looked at me, saying You do not believe what I have said, and I was forced to own that it was so.

For all his authority, dazzling skill, and mastery of his ship and men, FitzRoy was still only twenty-six years old, a young, imperious, seldom challenged or questioned aristocrat accustomed to getting his way, and this was a new view of him for the amiable Darwin. In the face of physical adversity, FitzRoy was uncowed, resourceful, and brave. But when he was thwarted in any personal way, a bratty petulance broke through his cool demeanor, and a darker side of his character took over. It might be a brief possession, and FitzRoy's natural grace and charm could quickly dispel it, but it came from a deep reservoir he would never escape.

The *Beagle*'s crew also grew bored and fractious during the long enforced delay. The ship was still anchored in Barnet Pool on Christmas Day, which brought Darwin some insight into the character of the English seaman.

The whole of it has been given up to revelry, at present there is not a sober man in the ship: King is obliged to perform the duty of sentry, the last sentinel came staggering below declaring he would no longer stand on duty, whereupon he is now in irons. . . . Wherever they may be, they claim Christmas day for themselves, & this they exclusively give up to drunkedness — that sole & never failing pleasure to which a sailor always looks forward to.

The longed-for break in the weather came the very next day, December 26, and the ship might have sailed except that most of the crew were either drunk or missing ashore. "The ship has been all day in state of anarchy," wrote Darwin. A number of the drunkards were chained in the hold, while AWOL crew members were rounded up ashore.

But the good weather held. The *Beagle*, under command of her master, Edward Chaffers, weighed anchors again at 11 A.M. on December 27. FitzRoy and Darwin celebrated the departure by lunching ashore in a tavern with Lieutenant Bartholomew Sulivan as the *Beagle* tacked out of Plymouth into the Channel. They ate mutton chops and drank champagne and joined the ship by boat outside the breakwater at 2 P.M. Immediately all sails were set and filled with a fair breeze. The *Beagle* scudded away down-Channel at 8 knots.

Darwin, perhaps buoyed by the excitement of the long-awaited departure, and views from the deck of England's green coastline fast slipping away off the starboard beam, felt fine through the afternoon and evening.

The great voyage, of such unimagined consequence, was begun.

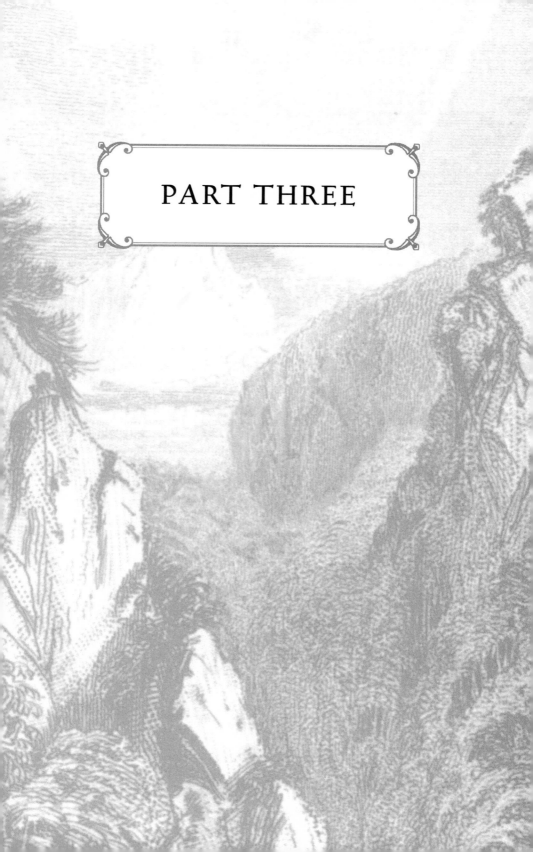

PART THREE

13

As the *Beagle* rolled and pitched across the Bay of Biscay, and then turned south into the open Atlantic, Darwin discovered that he was one of those few and unlucky voyagers who suffer from a chronic seasickness that does not get better the longer one is at sea. It was to plague him for five years.

"The misery is excessive & far exceeds what a person would suppose who had never been at sea more than a few days," he wrote in his diary on December 29, when the *Beagle* was 380 miles from Plymouth.

> I found the only relief to be in a horizontal position. . . . I often said before starting, that I had no doubt I should frequently repent of the whole undertaking, little did I think with what fervour I should do so. — I can scarcely conceive any more miserable state, than when such dark & gloomy thoughts are haunting the mind as have to day pursued me.

There was something else sickening him. As he lay in his swinging hammock fighting nausea, Darwin could not avoid hearing the lash and screaming of four seamen being flogged for drunkenness and disobedience on Christmas Day. FitzRoy noted the punishment in his captain's log of December 28, 1831:

John Bruce: 25 lashes for drunkenness, quarrelling and insolence.

David Russel: Carpenter's crew, with 34 lashes for breaking his leave and disobedience of orders.

James Phipps: 44 lashes for breaking his leave, drunkenness and insolence.

Elias Davis: 31 lashes for reported neglect of duty.

Darwin was appalled. FitzRoy justified such punishment to him in the terms he would later write in his journal: "Hating, abhorring corporal punishment, I am nevertheless fully aware that there are too many coarse natures which cannot be restrained without it, (to the degree required on board a ship,) not to have a thorough conviction that it could only be dispensed with, by sacrificing a great deal of discipline and consequent efficiency."

In a time and culture when men could not break the rigid barriers of rank and social class, when reasoning with a crew could be taken for weakness, this was the standard naval practice: discipline through the threat of severe punishment, a fundamental that was respected equally by officers and seamen. FitzRoy was no Bligh, but he was a strict martinet of the old school, which at times could seem like much the same thing.

Darwin's sudden immediate proximity to FitzRoy—eating with him daily, often accompanying him ashore—revealed a character that fascinated him as much as any natural phenomenon he encountered on his voyage around the world. His diary jottings, letters home, and passages from his autobiography provide history with the clearest observations of the mercurial young captain. "FitzRoy's character was a singular one," Darwin wrote years later,

with many very noble features; he was devoted to his duty, generous to a fault, bold, determined and indomitably energetic, and an ardent friend to all under his sway. He would undertake any sort of trouble to assist those whom he thought deserved assistance. He was a handsome man, strikingly like a gentleman with highly courteous manners. . . . FitzRoy's temper was a most unfortunate one, and was shown not only by passion, but by fits of long-continued moroseness against those who had offended him. His temper was usually worst in the early morning, and with his eagle eye he could generally detect something amiss about the ship, and was then unsparing in his blame. The Junior officers when they relieved each other in the forenoon used to ask "whether much hot coffee had been served out this morning?" which meant how was the captain's temper?

In London, on the day he had first met FitzRoy, Darwin had written a letter to his sister Susan expressing his initial enthusiasm for his "beau ideal" of a captain. Writing home from Brazil, he had to qualify this.

And now for the Captain, as I daresay you feel some interest in him. As far as I can judge, he is a very extraordinary person. I never before came across a man whom I could fancy being a Napoleon or a Nelson. I should not call him clever, yet I feel convinced nothing is too great or too high for him. His ascendancy over everybody is quite curious; the extent to which every officer and man feels the slightest praise or rebuke would have been before seeing him incomprehensible. . . . His candour and sincerity are to me unparalleled; and using his own words his "vanity and petulance" are nearly so. I have felt the effects of the latter. . . . His great fault as a companion is his austere silence produced from excessive thinking. His many good qualities are numerous: altogether he is the strongest marked character I ever fell in with.

But FitzRoy was a very good friend, after his own fashion, to Darwin. He encouraged his naturalist activities, putting his crew and ship and the facilities of the Royal Navy at Darwin's disposal. As the voyage wore on, Darwin's mounting collection—

boxes and barrels of plants and animals—were constantly shipped back to England, free of charge, by navy ships, under the direction of FitzRoy, with the blessing of the lords of the Admiralty.

More particularly, aboard the Beagle, FitzRoy and Darwin assumed the sort of respectful friendship Darwin had enjoyed with his peers at Cambridge. They called each other, in the manner of the English upper classes, by their surnames. They ate together, they found in each other a fellow scientist with whom to share findings and triumphs. FitzRoy was tireless in his efforts to make Darwin comfortable aboard the Beagle. With his own hands, the captain retied Darwin's hammock during their first days at sea. When Darwin later mentioned this kindness in a letter home, his father wept at such solicitude. FitzRoy soon began referring to Darwin as the "ship's philosopher"—since a naturalist was one who pursued the study of natural philosophy—and this quickly contracted to "Philos," as Darwin was affectionately called by FitzRoy and the entire crew.

This friendship with the captain was vital for Darwin, a supernumerary on a voyage that, in the long view, would be all about Darwin. But his singular position as a putative friend, someone who had been invited to express his views to the autocratic spring-wound twenty-six-year-old captain as an equal, would test that friendship to its breaking point.

Darwin had hoped for respite from his seasickness at Tenerife, where the Beagle was supposed to make its first stop. After closely reading about von Humboldt's travels around the island, he had dreamed of visiting Tenerife with his mentor Henslow. But "Oh misery, misery," he wrote in his diary: the local fear of cholera and overzealous quarantine regulations forbade anyone to go ashore for twelve days. FitzRoy wouldn't wait that long; he raised anchor immediately and sailed on. Darwin could only "gaze at this long-wished-for object of my ambition" from the deck. "Everything has a beautiful appearance: the colours are so rich and soft. The peak or sugar loaf has just shown itself above the clouds. It towers in the sky twice as high as I should have dreamed of looking for it"—and watch it fall below the horizon. On January 16th, three weeks out of England, the *Beagle* anchored in Porto Praya, on Saint Jago, one of the Cape Verde Islands. Darwin went ashore immediately and "strolled about the town, & feasted upon oranges."

Before returning to the ship, he walked beyond the small shantytown into a deep, unspoiled valley.

> Here I first saw the glory of tropical vegetation. Tamarinds, Bananas & Palms were flourishing at my feet. —I expected a good deal, for I had read Humboldts descriptions & I was afraid of disappointments: how utterly vain such fear is, none can tell but those who have experienced what I to day have. — It is not only the gracefulness of their forms or the novel richness of their colours, it is the numberless & confusing associations that rush together on the mind, & produce the effect. — I returned to the shore, treading on Volcanic rocks, hearing the notes of unknown birds, & seeing new insects fluttering about still newer flowers. — It has been for me a glorious day, like giving to a blind man eyes. — he is overwhelmed with what he sees & cannot justly comprehend it. — Such are my feelings, & such may they remain.

So Darwin wrote in his diary aboard the *Beagle* that evening. That day he found his voice. The next day, accompanying FitzRoy in one of the ship's boats to Quail Island, "a miserable, desolate" rock near Porto Praya, he found himself.

He had been reading Charles Lyell's first volume of *Principles of Geology* and looking at the rocks and sea pools around Quail Island when it occurred to Darwin that he might someday write his own book, one worth reading. Fifty years later he remembered the impact of this thought at that moment.

It then first dawned on me that I might perhaps write a book on the geology of the various countries visited, and this made me thrill with delight. This was a memorable hour to me, and how distinctly I can call to mind the low cliff of lava beneath which I rested, with the sun glaring hot, a few strange desert plants growing near, and with living corals in the tidal pools at my feet.

While FitzRoy busied himself ashore with surveying and problems of magnetic variation, Darwin tramped around Saint Jago for almost three weeks, alone or with friends from the ship's company, collecting everything that captured his fancy. He littered the *Beagle*'s deck and chart room with his specimens, causing First Lieutenant Wickham to complain, uselessly, about his mess. He examined, catalogued, and boxed everything for eventual shipment back to England, where Professor Henslow would oversee his growing collection.

The *Beagle* sailed from the Cape Verde Islands on February 8, heading for Bahia on the coast of Brazil, the first stop on FitzRoy's mission to enlarge the earlier surveys of South American waters. The ship bowled along before the trade winds on seas of a deep vivid blue unknown in colder, darker Europe. The water was warm, the air warmer and moist, the sky dotted with puffy cumulonimbus clouds. Rain squalls overtook the ship periodically, the sudden winds heeling her under a press of too much sail for ten or fifteen minutes, too short a time for the men to reduce canvas, so that the *Beagle* suddenly accelerated and rolled and made noisy, frothing waves that coursed past her hull until the cloudburst moved away leaving the deck dark and cool underfoot from the rain.

In the calmer tropical seas, Darwin felt better. He constructed a net and towed it astern on a long line, trawling for plankton and tiny sea creatures. He was able to work on his collection—dissecting plants and animals, writing up his notes—and settle into a shipboard routine.

He and FitzRoy met at eight every morning for breakfast in

the captain's cabin, again at 1 P.M. for dinner, and at 5 P.M. for supper. The first two meals were spartan, though Darwin found them ample and satisfying: rice, peas, bread, antiscorbutics like pickles and dried apples, water and coffee. For supper there was meat—from the cans while those supplies lasted, or fresh meat or fowl—bread and cheese. They drank no alcohol, by FitzRoy's preference. During these meals, the two men talked of their work when FitzRoy felt communicative, though often he did not, and ate in cogitative silence. They made a practice of leaving the table as soon as they were finished, without waiting for the other.

Twenty days from the Cape Verdes, the *Beagle* anchored in Bahia de Todos Santos (present-day Salvador). The town, "embosomed in a luxuriant wood," sent Darwin into raptures: "It would be difficult [to] imagine, before seeing the view, anything so magnificent . . . if faithfully represented in a picture, a feeling of distrust would be raised in the mind." He was soon spending his days wandering through the Brazilian forest, seeking each evening back aboard the *Beagle* adequate expression for what he had seen and felt.

29th (Feb)   The day has passed delightfully: delight is however a weak term for such transports of pleasure . . . amongst the multitude it is hard to say what set of objects is most striking . . . a most paradoxical mixture of sound & silence pervades the shady parts of the wood. — the noise from the insects is so loud in the evening it can be heard even in a vessel anchored several hundred yards from the shore. — Yet within the recesses of the forest when in the midst of it a universal stillness appears to reign. — To a person fond of Natural history such a day as this brings with it pleasure more acute than he ever may again experience. — After wandering about for some hours, I returned to the landing place. — Before reaching it I was overtaken by a Tropical storm.—I tried to find shelter under a tree so thick that it would never have been penetrated by common English rain, yet here in a couple of minutes, a little torrent flowed down the trunk. . . .

March 1st   I can only add raptures to the former raptures. . . .
Brazilian scenery is nothing more nor less than a view in the
Arabian Nights. . . .

But there was more than loveliness to contend with. At this
time the slave trade was still legal in Brazil, and Darwin, whose
ancestors had been strident abolitionists, was repelled by the stories told by Captain Paget of the *Samarang*, who came aboard
the *Beagle* to dine with FitzRoy.

Facts about slavery so revolting, that if I had read them in England, I should have placed them to the credulous zeal of well-
meaning people: The extent to which the trade is carried on; the
ferocity with which it is defended; the respectable (!) people
who are concerned in it are far from being exaggerated at home.
. . . It is utterly false (as Cap Paget satisfactorily proved) that
any, even the very best treated, do not wish to return to their
countries. — "If I could but see my father & my two sisters once
again, I should be happy. I can never forget them." Such was the
expression of one of these people.

But FitzRoy had different ideas. While he felt slavery was "an
evil long forseen and now severely felt," he believed the majority
of Brazilians treated their slaves humanely. He felt the institution
was not unlike the mutually useful master-servant, landowner-
tenant relations in existence in England since feudal times, long
enjoyed by his own family. He cited his own recent visit to a
Brazilian plantation where the owner had brought a number of
his slaves to meet the captain and asked them, in front of
FitzRoy, if they would rather be free. All had answered no.

Darwin, unable to restrain himself, grew uncharacteristically
angry and asked FitzRoy if he really thought the answers given
by slaves in the presence of their master were believable.

To FitzRoy, such a questioning of his opinion was almost
unknown; it was practically mutinous. He erupted furiously at

Darwin, saying that as he doubted his word, they could no longer "live together." The meal broke up instantly, and FitzRoy sent for Wickham to tell him that Darwin was no longer welcome at his table.

Darwin was convinced that his voyage was over, that he would have to leave the ship. But Wickham, perhaps more used to his captain's blacker moods, invited Darwin to take his meals with the officers in the gun room. It wasn't necessary: a few hours later FitzRoy sent an officer to him with an apology and a request that he "continue to live with" the captain. Darwin agreed, and they settled, without reference to the episode, back into their old routine. But the younger man was now aware of the terms of their friendship.

The *Beagle* weighed anchor and sailed from Bahia on March 15, her course southward.

The three Fuegians, who had sailed north along this same coast two years earlier, knew where they were headed.

14

They had come a remarkably long way. There was no more certain sign of the Fuegians' anglicization than their clothing. The first thing FitzRoy had done after kidnapping them, even before feeding them, was to dress them in English clothing. This, he assured the Admiralty in his first communication concerning his protégés, made them "very happy."

Now, two years on, they walked about the ship in their fancy duds and gazed shoreward from the decks, the oddest of tourists. There was no cruising or yachting gear for passengers, no white ducks or striped blazers. Though they may have been given oiled canvas seacoats to permit them to get some air on deck during inclement weather, the Fuegians would not have dressed as seamen. York Minster and Jemmy Button, in earnest collusion with their captors over their transformation, dressed in the *de rigueur* fashion of early-nineteenth-century gentlemen: aboard ship and ashore, they wore topcoats with tails, double-breasted waistcoats with lapels, high-collared shirts with cravats, long trousers, and leather boots. They had probably been given cheap watches with fobs to complete the proper "weskit" effect. The full regalia, pounds of English wool, must have been sweltering in the tropics, but it was the clear badge of their elevation from savagery and they wore every layer of it devoutly.

Jemmy Button in particular was observed by everyone to be fastidious about his dress. He had grown fat and vain during his stay in England, he was rarely seen without his white kid gloves, and was scrupulous, even neurotic, about the polish of his boots. While his speech never went far beyond the basic "Me go you" plateau of essential communication, he had an ear for the delicate and foppish in expression. When he visited Darwin in his seasick berth, Jemmy gazed at him with pity and said, "Poor, poor fellow!" This virtual satire of excessive Englishness—closely resembling someone from the lower classes putting on airs—amused captain, crew, and Darwin alike, and endeared Jemmy Button to all of them. He instinctually, if incompletely, understood this and played to his gallery.

Fuegia Basket's dresses and bonnets were probably more comfortable at sea and in the heat, but no less proper. And all three dressed up in their formal best for FitzRoy's regular Sunday shipboard services of hymn singing and prayers.

Although thrown into intimate contact with them for more than a year, Darwin's impressions of the Fuegians were less savvy than his observations of the natural landscapes he glimpsed at the *Beagle*'s ports of call. He agreed largely with FitzRoy's opinions of their innate personalities and characteristics, which the captain had derived from his ideas about facial features and the mumbo jumbo of phrenology. Soon after their arrival in England, FitzRoy had taken them to a phrenologist to have the bumps on their heads read. The specialist's finding was that all three were "disposed to cunning" and possessed "animal inclinations and passions" that would pose problems in making them "usefeul members of society." Nobody in Christian England would have been surprised by such a diagnosis. Darwin saw a gentler side to Jemmy Button, but his own conclusion that the boy must possess a "nice disposition" was based less on his day-to-day contact with him than his reading of the physiognomy of Jemmy's face.

Darwin thought the twelve-year-old Fuegia Basket "a nice, modest, reserved young girl, with a rather pleasing but sometimes sullen expression." He noticed too that she possessed an innate cleverness: "[She was] very quick in learning anything, especially languages. This was showed by picking up some Portuguese and Spanish, when left on shore for only a short time at Rio de Janeiro and Montevideo, and in her knowledge of English." Fuegia was the best English speaker of the three, and she picked up manners with her languages. She was the charmer, the pet of the lower deck on the voyage back to England from Tierra del Fuego, FitzRoy's showpiece specimen who had won Queen Adelaide's heart. Fuegia Basket had a natural charm that needed no translation, that bridged any cultural gap, an endearing quality as potent in its way as sex appeal. She understood this and used it.

Darwin's impressions of York Minster were close to FitzRoy's, for the little he wrote of the elder, most intractable Fuegian echoes the phrenologist's report, which Darwin undoubtedly read: "His disposition was reserved, taciturn, morose, and when excited violently passionate; his affections were very strong towards a few friends on board; his intellect good."

Darwin was most struck by the two Fuegian men's eyesight. His own had proved extremely sharp in shooting, and he had excellent distance vision, better than most of the crew's. But York and Jemmy sighted ships at sea and land beyond the horizon long before anyone else on board the ship. They were well aware of their superiority and enjoyed it: "Me see ship, me no tell," Jemmy liked to tease the officers on watch. Their sense of taste seemed keener too; they appeared to Darwin to have natural powers far beyond the capabilities of Europeans, an impression he was to amplify years later when he came to write his *Descent of Man*.

The clearest object of York Minster's affections was Fuegia Basket. It was understood by all on board that once landed in

Tierra del Fuego, York and Fuegia would become, as the sailors put it, "man and wife." Until then, while in his care, FitzRoy kept them apart as far as possible. Fuegia's hammock was swung aft, near the officers' quarters, while Jemmy and York bunked forward with the crew.

It was a long slow cruise south. The *Beagle* made lengthy stops in Rio de Janeiro (FitzRoy left Fuegia Basket ashore here in the company of an English family for three months, where she helped the young children of the family with their English, while picking up Portuguese herself), Montevideo, and Bahia Blanca on the Argentine coast. Between these ports the ship cruised painstakingly back and forth surveying the coast.

Darwin took advantage of this tedious cruising to explore ashore for weeks at a time, renting houses and staying with ranchers, wandering through Brazilian rain forests, galloping across the Argentine pampas with bands of gauchos, thrilling and exhausting himself. He also spent more time in Buenos Aires and Montevideo. For reading material ashore, surrounded by paradisaical natural glories, he carried Milton's *Paradise Lost* everywhere. He stayed in touch with the ship and its wanderings by a fairly regular correspondence with FitzRoy. The tone of these letters shows the essential warmth of their friendship. The captain clearly missed Darwin while he was off the ship; his letters reveal an antic, schoolboy banter—a tone he could never have enjoyed with his subordinates—underscoring both FitzRoy's tender age and the lonely isolation of his rank and often fearsome responsibility:

> My dear Philos,
>   Trusting that you are not entirely expended — though half-starved, occasionally frozen and at times half drowned — I wish you joy from your campaign with General Rosas [an

Argentine general whose troops were slaughtering Indians all over the pampas], and I do assure you that whenever the ship pitches (which is *very* often as you *well* know), I am extremely vexed to think how much *sea practice* you are losing; — and how unhappy you must feel on firm ground.

Your home (upon the waters) will remain at anchor near the Monte Megatherii until you return to assist in parturition of a Megalonyx measuring seventy-two feet from the end of his snout to the tip of his tail, and an Ichthyosaurus somewhat larger than the *Beagle*! . . . [While still with the ship in Bahia Blanca, Darwin and FitzRoy had discovered on a nearby beach the fossilized bones of several large animals, which they believed could have belonged to a *Megatherium* or *Megalonyx*, both extinct.]

My dear Darwin,

Two hours since I received your epistle. . . . and most punctually and immediately am I about to answer your queries. (Mirabilo!!) But firstly of the first — My good Philos, why have you told me nothing of your hairbreadth escapes and moving accidents? How many times did you flee from the Indians? How many precipices did you fall over? How many bogs did you fall into? How often were you carried away by the floods? . . . I hear you are saying, "You have got to the end of a sheet of paper without telling me one thing that I want to know."

Philos, do not be irate, have patience and I will tell thee all.

Tomorrow we shall sail for Maldonado — there we shall remain until the middle of this month — thence we shall return to Monte Video . . .

Adios Philos — Ever faithfully yours,
Robt. FitzRoy

FitzRoy was anxious to get south to Tierra del Fuego to land the Fuegians and Matthews, the missionary, and help them establish their mission during the brief southern summer, December to January. In September 1832, daunted by the enormity and difficulty of making a thorough survey of the Argentine coast with the *Beagle* alone, he hired two small local sealing schooners to share some of the work. The seventeen-ton *Paz* was, FitzRoy wrote, "as ugly and ill-built a craft as I ever saw, covered with dirt, and soaked with rancid oil." The eleven-ton *Liebre* was just as filthy, but they appeared seaworthy and suitable. FitzRoy outfitted them from the *Beagle's* stores, manned them with his own men, and sent them off to survey the shoal waters and river estuaries between Bahia Blanca and Rio Negro.

By early December, the survey work was done. The *Beagle* returned to Montevideo for supplies, and then sailed south again.

On December 16, 1832, the *Beagle* closed with the bleak eastern shore of Tierra del Fuego. Through the long afternoon and twilight the ship sailed southeast, paralleling the shore at a distance of a few miles, and in the evening anchored in an exposed bight off Cape Santa Inez, some 130 miles south of the entrance to the Strait of Magellan. The land ran in a long unbroken line northwest to southeast, offering no safe harbor, ending in sheer cliffs over which seabirds wheeled and cried.

"The sky was gloomy," wrote Darwin in his diary. "At a great distance to the south was a chain of lofty mountains, the summits of which glittered with snow."

The *Beagle* had never visited this part of the Fuegian coast. The crew wondered if it was inhabited. Shortly they knew: smoke rose from the shore. Through his telescope, Darwin saw Indians scattered about the sheer edge of the land "watching the ship with interest." This southeastern shore of Tierra del Fuego was the territory of the Ona tribes.

With their keen eyesight, the Fuegians aboard the *Beagle* did not need telescopes. "Oens-men—very bad!" Jemmy and York told FitzRoy, and asked him to fire at them, which he declined to do.

Despite his refusal to kill their enemies, FitzRoy observed that the three Fuegians appeared elated. They stared shoreward from the deck and knew they were home.

The captain's own feelings at seeing "his" Fuegians so close to their repatriation went unrecorded.

15

Nowhere are the designs of men more subject to the cooperation of the natural world than among the dangers of the Cape Horn region. All of FitzRoy's hopes and plans, his schemes for the Fuegians, his surveying mission, the lives of all aboard his ship, depended on his carrying them all through a war zone of wind and waves. Above and beneath every other piece of business now came the daily struggle to survive conditions that had long made of this place a graveyard.

Almost immediately, this meant flight. A heavy swell, harbinger of an Atlantic storm far to the north, came in the night and set the *Beagle* rolling violently. There was little breeze, but as the swell deepened, FitzRoy, who had hoped to spend a day or two in the vicinity to make observations, grew anxious that the waves and the wind that might follow them could drive the ship into the cliffs. At 3 A.M., when first light came, he gave the order to weigh anchor, but with her tethers to the seafloor raised, the ship drifted uncertainly, pushed shoreward by the swell as the crew scrambled in the rigging to work her seaward in light airs. Then the breeze FitzRoy had anticipated sprang up, filled the sails, and carried the *Beagle* out of immediate danger. But weather was coming and there was no protection from it on this long inhospitable shore, so FitzRoy turned his ship south again and ran for the Strait of Le Maire.

At noon, the masthead lookout reported very high breakers ahead off Cape San Diego: the Southern Ocean's flood tide was pouring east from Cape Horn like a millrace through the strait, piling up against the rising north wind and swell, creating seas that could overwhelm a modern tanker. "The motion from such a sea is very disagreeable," wrote Darwin with considerable understatement; "it is called 'pot-boiling,' & as water boiling breaks irregularly over the ship's side." The *Beagle* escaped a pot-boiling that day for when she reached Cape San Diego an hour later at 1 P.M., the tide had turned, running now in the same direction as the wind and waves; the breakers disappeared, and the *Beagle* sluiced through the strait on the new, and now favorable, ebb tide.

A few hours later, close under the lee of the Fuegian shore, the ship sailed into Good Success Bay (any anchorage in the Strait of Le Maire is a success story), and Darwin got his first look at Fuegians not in English clothing, but in their natural element.

> In doubling the Northern entrance [to Good Success Bay], a party of Fuegians were watching us, they were perched on a wild peak overhanging the sea & surrounded by wood. — As we passed by they all sprang up & waving their cloaks of skins sent forth a loud and sonorous shout. — this they continued for a long time. — These people followed the ship up the harbor & just before dark we again heard their cry & soon saw their fire at the entrance of the Wigwam which they built for the night.

As soon as the *Beagle*'s anchors were down, the wind shifted from the north to the southwest and began to blow hard. Heavy squalls and hurricane-force williwaws swept down upon the ship from the high surrounding hills, but the water in the anchorage remained flat, undisturbed by waves or swell, and the fine white sand on the floor of Good Success Bay proved to be firm holding ground.

This was Darwin's first experience of Cape Horn conditions, but he was confident of the ship and her men.

Those who know the comfortable feeling of hearing rain & wind beating against the windows whilst seated around a fire, will understand our feelings: it would have been a very bad night out at sea, & we as well as others may call this Good Success Bay.

The next morning Darwin met a group of Fuegians ashore. He was amazed: "I would not have believed how entire the difference between savage and civilized man is," he wrote in his diary.

With just the skin of a guanaco (a South American animal similar to a llama) thrown over their shoulders, daubed with red and white "paint" and charcoal, they reminded Darwin strongly of the Wolf's Glen devils in Carl Maria von Weber's gothic *Der Freischutz*, which he had seen on the stage in Edinburgh eight years earlier.

The elder "devil" slapped him simultaneously on the chest and back three times while making "the same noise which people do when feeding chickens," and then asked Darwin to slap him back in the same fashion. He obliged, making the Fuegian happy. Darwin concurred with Captain Cook's description of the sound of the Fuegian language.

It is like a man trying to clear his throat; to which may be added another very hoarse man trying to shout and a third encouraging a horse with that peculiar noise which is made in one side of the mouth. Imagine these sounds and a few gutturals mingled with them, and there will be as near an approximation to their language as any European may expect to obtain.

He was equally struck by the primitiveness of their lifestyle.

If their dress and appearance is miserable, their manner of living is still more so. — Their food chiefly consists in limpets & mussels, together with seals & a few birds; they must also catch occasionally a Guanaco. They seem to have no property excepting bows & arrows & spears: their present residence is under a few bushes by a ledge or rock: it is no ways sufficient to keep out rain or wind. — & now in the middle of summer it daily rains & as yet each day there has been some sleet. — The almost impenetrable wood reaches down to the high water mark. — so that the habitable land is literally reduced to the large stones on the beach. — & here at low water, whether it be day or night, these wretched looking beings pick up a livelihood. — I believe if the world was searched, no lower grade of man could be found. — The Southsea Islanders are civilized compared to them, & the Esquimaux, in subterranean huts may enjoy some of the comforts of life.

He also observed, as others had before him, that the Fuegians were excellent mimics, able to mime the Englishmen's physical mannerisms precisely and parrot back whole sentences in the English they couldn't understand. What European could possibly do that, or follow an American Indian through a sentence of more than three words? Darwin wondered. He believed this skill was a complement of the superior natural powers of taste, smell, and sight that he had observed in the Fuegians aboard the *Beagle*.

He returned to the ship for lunch and was rowed ashore again that afternoon with a party that included FitzRoy, Jemmy Button, and York Minster.

This was FitzRoy's first encounter with Fuegians in their natural state since leaving Tierra del Fuego two years earlier. He was impressed all over again by their savagery, but now he saw that condition, with a kind of brotherly affection, as improvable.

Disagreeable, indeed painful, as is even the mental contemplation of a savage, and unwilling as we may be to consider ourselves even remotely descended from human beings in such a state, the reflection that Caesar found the Britons painted and clothed in skins, like these Fuegians, cannot fail to augment an interest excited by their childish ignorance of matters familiar to civilized man, and by their healthy, independent state of existence.

In attempting to describe their color, FitzRoy sought a palette of comparisons.

> A rich reddish-brown, between that of rusty iron and clean copper, rather darker than copper, yet not so dark as good, old mahogany. . . . The colour of these aborigines is extremely like that of the Devonshire breed of cattle. From the window of a room in which I am sitting, I see some oxen of that breed passing through the outskirts of a wood, and the partial glimpses caught of them remind me strongly of the South American red men.

The two "civilized" Fuegian men, Jemmy and York, became instant snobs. They alternately laughed at the squalor of their countrymen and appeared ashamed by them. They even pretended not to understand the Fuegians' speech (for Jemmy this might have been true), but York Minster could not help laughing hysterically when one of the older Fuegians, recognizing him as a fellow native despite his fine clothes, chided him and told him he was dirty for not shaving the few hairs on his face.

Richard Matthews, the missionary who was to establish a new Jerusalem on these wild shores, was also getting a first look at his future parishioners in the raw. His true impressions were not recorded, but according to FitzRoy he remained stoically unfazed: "[he] did not appear to be at all discouraged by a close inspection of these natives. He remarked to me, that 'they were no worse than he had supposed them to be.'"

After three days of survey work, while Darwin and others climbed the nearby hills and attempted unsuccessfully to shoot one of the giant guanacos they had spotted, the *Beagle* sailed from Good Success Bay. FitzRoy was eager to settle Matthews and the Fuegians ashore at last and tried to make for Christmas Sound and March Harbour, west of Cape Horn—the neighborhood where his fine whaleboat had been stolen two and a half years earlier, and where he had abducted York Minster and Fuegia Basket.

The ship passed south of the Horn on December 21, but then the wind shifted and began to blow at gale force, as it so often does here, driving them out to sea. Two days later, on Christmas Eve, the *Beagle*'s crew worked her through driving hail into a small cove on Hermite Island, just west of Cape Horn. There they remained for Christmas and until the end of the month, in a secure anchorage, while storms and williwaws blew around them.

On December 31, though conditions were little improved, the *Beagle* weighed anchor, as FitzRoy was impatient to reach March Harbour, now only 100 miles away. But the weather remained relentlessly against him. For the first two weeks of January he pushed the *Beagle* westward through a succession of gales, battling the entire time to make that short distance. For days the ship made no headway at all: the tiny offshore Diego Ramirez Islands were sighted through the murk close off the ship's port beam on January 2 and again in exactly the same place on January 5. Life aboard the ship beating with no letup into gale force winds and icy breaking seas was reduced to the grimmest continuum of food, water, rest, and struggle. Seawater made its way everywhere below, through hatches and streaming from the men's soaked clothing. This is what the *Bounty*—much the same size and shape as the *Beagle*—faced in this same spot in 1788, before Bligh gave up, turned at last downwind, and sailed around the world the other way to reach the Pacific.

The *Beagle*'s captain and crew were not as unhappy as the *Bounty*'s, but her natural philosopher was as miserable as he had ever been. Swinging wildly in his hammock, he complained to his diary: "[Since December 21] I have scarcely for an hour been quite free from seasickness. How long the bad weather may last,

I know not; but my spirits, temper, and stomach, I am well assured, will not hold out much longer."

On January 11, the towering rock formation that had given York Minster his name was sighted ahead, "looming among driving clouds." FitzRoy believed they would soon be at anchor in March Harbour, then only a mile ahead, when the gale suddenly increased to storm force and darkness, violent squalls, rain and hail drove the *Beagle* out to sea again.

All the following day the ship lay hove-to south and just west of Cape Horn, drifting slowly back over the sea miles she had won with so much effort. It was twenty-four days since they had passed the Horn and they were now barely twenty miles west of it. The storm steadily worsened until it reached a pitch of screaming intensity around noon on the 13th. It was the worst weather FitzRoy had ever encountered. The waves had grown to such heights that he remained on deck in the driving wind and rain, able to do nothing but watch them anxiously, feeling a sense of imminent catastrophe. At 1 P.M., three great rollers bore down on the ship.

[Their] size and steepness at once told me that our sea-boat, good as she was, would be sorely tried. Having steerage way, the vessel met and rose over the first unharmed, but, of course, her way was checked; the second deadened her way completely, throwing her off the wind; and the third great sea, taking her right a-beam, turned her so far over, that all the lee bulwark, from the cat-head to the stern davit, was two or three feet under water.

In other words, the ship was knocked right over on her side, capsized.

Water burst open doors and hatches, cataracts tumbled below into the chart room where Darwin lay in his misery and spread through the cabins. The *Beagle* tried to rise but wallowed on her side, listing with the weight of the water now trapped against her bulwarks. "Had another sea then struck her," wrote FitzRoy, "the little ship might have been numbered among the many of her class which have disappeared."

Lieutenant Sulivan struggled up onto the deck from below (as he later described the event to his son) and found the *Beagle* on her side. Carpenter Jonathan May was already sliding along the nearly vertical wall of the deck, struggling with a hand spike to open the hinging wooden ports at the edge—now the bottom—of the deck to allow the water to drain off. Sulivan helped him and when a few of these had been knocked open, the water drained off the deck and the ship came up. One of the new whaleboats had been torn off its davits astern and was hanging smashed alongside the hull; crewmen had to take an ax to its tackle to chop it adrift. But the ship was otherwise unscathed. None of its rigging had carried away, no one had been swept overboard. Apart from the whaleboat, the only recorded damage was to one chronometer and Darwin's "irreparable loss" of some of his specimens.

The knockdown marked the height of the storm. The wind soon abated sufficiently for the crew to set some sail and the ship turned north, off the wind, to find shelter. It was after dark when the *Beagle* let go her anchors in quiet water behind False Cape Horn at the edge of Hardy Peninsula, only twelve miles from her last anchorage on Hermite Island, which she had left two weeks earlier.

FitzRoy abandoned the attempt to reach March Harbour in the *Beagle.* Days of continued bad weather forced them north and east through Nassau Bay, still farther away, until he finally decided to leave the ship in Goree Road, a secure and accessible anchorage atop Nassau Bay, and proceed in the smaller boats through the inside passages among the islands to the area around Christmas Sound.

But York Minster stopped him. He told the captain he didn't

want to return to his own country. He and Fuegia, his wife to be, preferred to settle with Jemmy Button and Matthews in Jemmy's "country," York told the captain. FitzRoy was surprised but glad. He thought it much better that the three of them and Matthews all settle together.

"I little thought," he later wrote, "how deep a scheme Master York had in contemplation."

16

Darwin was among the large group that set out on January 19, headed for the place near Murray Narrows that Jemmy Button called his "conetree," from where he had first been taken. This was where FitzRoy now decided the new settlement and mission should be established.

For about a day, he had considered the land on Lennox and Navarin Islands around the *Beagle*'s anchorage in Goree Road. More towns and cities have come into being because of their proximity to a safe harbor than for any other reason. The coast hereabouts was unusually flat for this drowned-mountain region, and FitzRoy thought it might prove suitable to agriculture. But after tramping across it for a few miles with Darwin, they found the whole area to be a swamp, "a dreary morass . . . quite unfit for our purposes." Jemmy's country, where he was keen to return, seemed more promising.

FitzRoy, Darwin, Matthews and the three Fuegians, Bynoe, the *Beagle*'s surgeon, and about twenty-seven seamen, marines, and officers, filled three whaleboats and the ship's 26-foot yawl. Temporarily decked over, the yawl was also crammed with the cargo largely donated by the Church Missionary Society and others to start the Fuegians and Matthews on their new life.

The choice of articles [wrote Darwin] showed the most culpable folly & negligence. Wine glasses, butter-bolts, tea-trays, soup turins, [a] mahogany dressing case, fine white linen, beavor hats & an endless variety of similar things shows how little was thought about the country where they were going to. The means absolutely wasted on such things would have purchased an immense stock of really useful articles.

The folly and the culpability were mainly FitzRoy's, as Darwin surely knew. The captain, better than any of the well-meaning donors, knew the conditions of the wild, uttermost shores upon which the pioneers would be deposited. He had apparently not instructed the generous suppliers on what might have been more useful, nor sold or traded the linens, tea trays, and beaver hats for tools. He had been the prop master organizing the shipping and handling of all these silly items. He had watched them being carried into the *Beagle*'s hold at Devonport, where they had taken up valuable cargo space. It was one thing to have the queen of England presenting Fuegia Basket with a ring as a keepsake from a famous patron, but something else altogether to transport a boatload of genteel and mainly useless bric-a-brac to Tierra del Fuego. What could he have been thinking?

Here was the crazy myopia of FitzRoy's vision, fueled and abetted by Britain's expansionist aspirations and its own special relationship with God: three little wigwams raised against the williwaws of Cape Horn, each fixed up inside as an English drawing room, in which York, Jemmy, and Fuegia would repose, in their tailcoats and cravats and bonnet, with Matthews in the third wigwam to lead them in Sunday services. If such a thing was possible, what might not have grown from it? This was FitzRoy's dream, and he had pursued it with the energy of Alexander.

The boats traveled north from Goree Road and entered the Beagle Channel. This remarkable, two- to three-mile-wide, nearly straight, easily navigable waterway running east-west for 120 miles between high mountains, had been discovered by Murray, the *Beagle*'s master on the previous voyage.

Darwin's doubts about the venture didn't prevent him from enjoying the outing or the constantly magnificent scenery.

> In our little fleet we glided along, till we found in the evening a corner snugly concealed by small islands. — Here we pitched out tents & lighted our fires. — nothing could look more romantic than this scene. — the glassy water of the cove & the boats at anchor; the tents supported by the oars & the smoke curling up the wooded valley formed a picture of quiet & retirement.

It was like a camping scene by the American painter-franchiser Thomas Kincaid, whose lurid sublimity and mossy palette would be just right for the exaggerated picturesqueness of Tierra del Fuego.

As the group pulled west along the channel the next day, Fuegians appeared on the shore. They gaped at the cruisers in astonishment and ran for miles beside the channel to keep up with the boats. The Beagle Channel does not appear to have been noticed by voyagers before Murray came upon it in 1830; it would have been out of the way and unknown to the sealing and whaling vessels that frequented the region's ocean coasts or the Strait of Magellan, so it's probable that many of the Fuegians who saw the Englishmen in their boats that day had never seen any but their own people before. Fires sprang up all along the coast, both to attract the strangers' attention and to spread the news of their presence. For Darwin, the Fuegians delivered another gothic-opera spectacle:

> I shall never forget how savage & wild one group was. — Four or five men suddenly appeared on a cliff near us. — they were absolutely naked & with long streaming hair; springing from the ground & waving their arms around their heads, they sent forth the most hideous yells. Their appearance was so strange, that it was scarcely like that of earthly inhabitants.

The Anglicized Fuegians traveling with the Englishmen thought no better of these people than the group they had met a few weeks earlier in Good Success Bay. "Large monkeys," York Minster called them, laughing at them with what to FitzRoy must have been a dispiriting lack of Christian feeling. (Since there were no monkeys in Tierra del Fuego, York could only have acquired this derogatory comparison in England, and it's not hard to imagine how.) Jemmy Button assured the captain that these people were greatly inferior to his own, who were "very good and very clean."

It was Fuegia Basket who had the strongest reaction to the sight of her countrymen in their original state—her first in two and a half years, since she hadn't come ashore with the others at Good Success Bay. She was plainly terrified. After two years of total immersion among the most fragrant of English sensibilities, she was shocked at their nakedness and brute appearance. She may also have felt an acute embarrassment: this was who and what she really was. She had been lifted out of this, taken to another world and given the profoundest makeover. But now she was being returned to starkest heathendom—it is nowhere recorded whether she was pleased to come back or not—to be left here with a chest full of dresses and petticoats, some tea trays, and her great hulking twenty-eight-year-old suitor York Minster. Fuegia was still only twelve years old at the most, an intuitive, clever girl to be sure. But such attributes may not have helped her at this moment, when dumb ignorance might have been preferable. Her grasp of all that had happened to her, and was about to, can only be guessed at. FitzRoy, with his tunnel vision, so ready to believe what he wanted to believe, is our only witness to her feelings. He wrote: "Fuegia was shocked and ashamed; she hid herself, and would not look at them [the wailing wild men ashore] a second time."

Two days farther westward along the Beagle Channel, the convoy passed below what is now the port of Ushuaia. That evening, in a cove at the north end of the Murray Narrows, they met a small group of "Tekeenica" Fuegians. They were members of Jemmy Button's tribe, whom he remembered, and they remembered him.

Both FitzRoy, and therefore Darwin and others, described Jemmy's tribe as the Yapoo Tekeenica, or the Yapoo division of the Tekeenica tribe. It's not clear where FitzRoy got this, though he wrote that he believed he had heard Boat Memory and York Minster referring to the Fuegians of this area as Yapoos on the previous voyage. But he was mistaken. The tribe, and Jemmy Button, actually called themselves Yamana. Lucas Bridges, son of the missionary Thomas Bridges, who later established a mission at Ushuaia and produced a 32,000-word Yamana-English dictionary, explained such a misunderstanding in his book *Uttermost Part of the Earth*.

It is interesting to note how many names have arisen through mistakes and even become permanent by finding their way into Admiralty charts. Early historians tell us of a place called Yaapooh, and speak of the people of that country. No such place or people existed, and this word is simply a corruption of the Yaghan [Yamana] name for otter, *iapooh*. No doubt FitzRoy, pointing towards a distant shore, asked what it was called. The [Fuegian's] keen eyes would spy an otter, and he would answer with the word, "Iapooh."

In all the charts of this country—both Spanish and English— a certain sound in Hoste Island bears the name Tekenika. The Indians had no such name for that or any other place, but the word in the Yaghan tongue means "difficult or awkward to see or understand." No doubt the bay was pointed out to a native, who, when asked the name of it, answered, "Teke uneka," implying, "I don't understand what you mean," and down went the name "Tekenika."

Much of what FitzRoy "learned" from his dealings with Fuegians must be set against misunderstandings like this—and his reliance on translators who were frequently under coercion.

Although they were his people, Jemmy found he had forgotten much of his native language and had trouble understanding them. York Minster, though from another tribe, did better and acted as a translator. In this way, Jemmy heard the news that his father had died. He had already had a "dream in his head" to that effect, wrote Darwin, so he seemed unsurprised, but according to FitzRoy, he looked "grave" at this news, went and found some green branches, which he burned with a solemn look. After that, he returned to his usual, cheerful self.

In the morning, a large number of natives arrived at the cove as the Englishmen were breaking camp. Many had run so fast over the mountains from Woollya (now Wulaia) that blood was streaming from their noses. Their mouths foamed as they talked, feverishly lobbing questions at Jemmy and the others. Bleeding, frothing at the mouth, gasping for breath, painted white, red, and black, they looked like "so many demoniacs" according to Darwin.

These were all "Tekeenicas," natives of southeastern Tierra del Fuego according to FitzRoy, who believed he saw marked differences between them and other Fuegians. These were

> low in stature, ill-looking, and badly proportioned. Their colour is that of very old mahogany, or rather between dark copper, and bronze. The trunk of the body is large, in proportion to their cramped and rather crooked limbs. Their rough, coarse, and extremely dirty black hair half hides yet heightens a villanous expression of the worst description of savage features.

This is FitzRoy's standard description of all Fuegians in the wild, no matter where he saw them. It was how he saw the ungodly savage anywhere. It fitted not only his own drawing of Jemmy Button, a "Tekeenica," after the ameliorating influences of his stay in England had worn off and left his features coarsened as of old, but also his later drawings of Maoris in New Zealand, whose lips curl threateningly and whose expressions are uniformly villainous. "Satires upon mankind" was FitzRoy's summing up of the physiognomy of Tekeenica men. He was no more generous with the women.

> They are short, with bodies largely out of proportion to their height; their features, especially those of the old, are scarcely less disagreeable than the repulsive ones of these men. About four feet and some inches is the stature of these she-Fuegians— by courtesy called women.

As the Englishmen's convoy got underway, they were joined by more natives on the water.

> In a very short time there were thirty or forty canoes in our train, each full of natives, each with a column of blue smoke rising from the fire amidships, and almost all the men in them shouting at the full power of their deep sonorous voices. As we pursued a winding course around the bases of high rocks or between islets covered with wood, continual additions were made to our attendents; and the day being very fine, without a breeze to ruffle the water, it was a scene which carried one's thoughts to the South Sea Islands, but in Tierra del Fuego almost appeared like a dream.

So this flotilla passed through the Murray Narrows to Woollya, the place where Jemmy Button had first been taken from a canoe. As they reached Ponsonby Sound at the southern end of the narrows, Jemmy recognized where he was. He now guided the boats into the quiet cove where he had once lived. There were only a few natives ashore. The women ran away, the remaining men nervously watched the boats land.

FitzRoy was immediately happy with the look of the place.

> Rising gently from the waterside, there [were] considerable spaces of clear pasture land, well-watered by brooks, and backed by hills

of moderate height, where we afterwards found woods of the finest timber trees in the country. Rich grass and some beautiful flowers, which none of us had ever seen, pleased us when we landed, and augured well for the growth of our garden seeds.

The English sailors pulled hard to stay ahead of the following canoes. As soon as they were ashore, the marines marked a boundary line on the ground with spades and spaced themselves out to guard the enclosed site on which the seamen now set to work erecting the settlement's wigwams and digging a garden. The canoes began arriving, and more natives gathered on the shore. York and Jemmy were kept busy explaining to them the meaning of the boundary line, and what was happening, and the natives squatted down to watch.

Woollya, just below Murray Narrows, Ponsonby Sound. The site of FitzRoy's, and others', hoped-for New Jerusalem in Tierra del Fuego.
(Narrative of HMS *Adventure* and *Beagle, by Robert FitzRoy*)

In the evening, a deep, booming voice was heard around the cove, coming from a canoe far down the sound. Jemmy recognized it instantly: "My brother!" he said. He abandoned the nails and tools he had been distributing, and scrambled onto a large rock to watch the canoe approach. It held, along with his stentorian brother, three younger brothers, two sisters, and his mother. The canoe was a long time in reaching the cove, and when it arrived, the family reunion was a strange one. His mother and sisters barely looked at him before running off to hide, Fuegian fashion. The brothers approached Jemmy slowly, their once naked brother now returned as fancy and particular about his dress as Phileas Fogg, and circled him wordlessly, like dogs sniffing a stranger. Jemmy stole glances at his English friends and suffered a mortification known in all cultures: he was embarrassed by his family. But the Englishmen were delighted, and the family immediately became "The Buttons," Jemmy's two older brothers becoming Tommy and Harry (in some accounts Billy) Button. At last Jemmy tried to speak to them. Darwin observed this meeting.

> It was pitiable, but laughable, to hear him talk to his brother in English & ask him in Spanish whether he understood it. I do not suppose, any person exists with such a small stock of language as poor Jemmy, his own language forgotten, & his English ornamented with a few Spanish words, almost unintelligible.

The thronging natives left the English camp at sunset, to set up their own fires and wigwams a quarter of a mile away. During the evening, Jemmy spent time with his mother and family, and York and Fuegia went visiting from wigwam to wigwam, explaining their presence, and the Englishmen's intention of establishing a settlement at Woollya. This seemed to have a calming effect on the locals, who appeared more relaxed the next day.

For the next four days, until January 27, the Englishmen worked at preparing the settlement that Matthews, York, Fuegia, and Jemmy were to call home. They gave them the best the Royal Navy and missionary zeal could provide. The sailors erected three homes. These were called wigwams, fashioned, like the native enclosures, of saplings and thatched with grass and twigs, but probably also wrapped with sailcloth and girded and strengthened with rope. Built by the ship's carpenters, riggers, sailmakers, bosuns, and seamen used to arranging the ingenious mechanical devices aboard a ship to their liking, they were substantial structures, built to last as long as possible, far superior to the makeshift, temporary, and transportable wigwams of the natives. Matthews's new home had both an attic made with boards to house his abundant stores, and a "cellar"—a pit beneath the floorboards—to secrete his more valuable possessions.

Near the wigwams, the seamen stepped off a good-sized plot and dug, planted, and sowed a kitchen garden of potatoes, carrots, turnips, beans, peas, lettuce, onions, leeks, and cabbages. The British were then a nation of gardeners—farmers, crofters, fruit and flower growers. Apart from its fast-diminishing forests, Britain was almost entirely under cultivation, and a green thumb lay dormant or active in every Briton. These *Beagle* gardeners probably longed, as seamen chronically do, for a home and garden of their own, or had them, far away, tended by a wife or family member. Throughout the eighteenth and nineteenth centuries, the British Navy and merchant service were often the only possibility of employment for young men from an expanding and overpopulated rural labor force who sought a living on farms or from their own smallholdings, whose only alternative was to look for work in densely overcrowded cities. The loving care these seamen expended on creating this English country garden for a few savages in remote Tierra del Fuego cannot be overestimated; the exchange of tips and advice coming from grizzled salts resembling the *Bounty* mutineers would have been as dedicated and earnest, authoritative and argumentative, as any gathering of the Royal Horticultural Society.

Around this industry, Fuegians continued to gather and watch.

Over the next few days their numbers grew to more than 300. Interaction between them and the Englishmen was initially harmonious, but there were constant attempts at thievery and incessant importuning with the word the English transliterated as "Yammerschooner." (According to Thomas Bridges's Yamana-English dictionary, *yamask-una* means "Do be liberal to me.") "The last & first word is sure to be 'Yammerschooner,'" wrote Darwin. The Englishmen gave them small presents but these were never enough. "It is very easy to please but as difficult to make them content. . . . they asked for everything they saw & stole what they could."

After several days, the pilfering grew bolder, and on January 26 there occurred several hostile incidents when a few older Fuegian men tried to force their way into the English encampment. One of them, rebuffed by a sentry from the site boundary, spat in the seaman's face and then pantomimed killing, skinning, and cutting up a man. FitzRoy was already concerned that his increasingly outnumbered party of thirty-odd men could be overwhelmed if the mood turned sour—the death of Captain Cook at the hands of his former devotees in Hawaii in 1779 cast a long shadow over subsequent relations between English seamen and aboriginal natives—so that evening he set the marines to some target practice with their muskets. The Fuegians watched this keenly, squatting on their haunches around the boundary like spectators at a fireworks display. FitzRoy had the targets arranged "so that they could see the effects of the balls." The natives were duly impressed, and afterward went off to their own camps, "looking very grave and talking earnestly."

The next morning, as the final thatching went into the wigwams, nearly all the Fuegians, including the Buttons, broke up their camps and paddled away or disappeared over the surrounding hills. Only half a dozen men were still too curious to leave. The English wondered if they'd been frightened off by the target practice, or whether an attack was being planned. FitzRoy decided to avoid any possibility of a conflict by withdrawing his men and marines to another cove a few miles away. Rather boldly, he also decided to leave Matthews and his three Fuegians to spend their first night—unguarded, with all their goods and stores—in the new wigwams. York and Jemmy both told FitzRoy that they were sure they would come to no harm, and Matthews appeared as steady and trusting as ever. The captain was impressed by his stoicism. At sunset, the four boats paddled away, leaving the settlers behind.

FitzRoy passed a sleepless night.

> I could not help being exceedingly anxious about Matthews, and early next morning our boats were again steered towards Woollya. My own anxiety was increased by hearing the remarks made from time to time by the rest of the party, some of whom thought we should not again see him alive; and it was with no slight joy that I caught sight of him, as my boat rounded a point of land, carrying a kettle to the fire near his wigwam. We landed and ascertained that nothing had occurred to damp his spirits, or in any way check his inclination to make a fair trial. Some natives had returned to the place, among them one of Jemmy's brothers; but so far were they from showing the slightest ill-will, that nothing could be more friendly than their behaviour.

Since all seemed well, FitzRoy decided to leave the group in Woollya for another week or so while he explored the western arms of the Beagle Channel, and then return to see how they were doing. He sent the yawl and one whaleboat back to the *Beagle* in Goree Road and set out with Darwin and a smaller group in the other two boats.

For a week, Darwin and FitzRoy had exciting but relatively trouble-free cruising. The weather began hot and sunny, and they were surprised to find themselves sunburned. They saw many whales breaching and spouting in the channel, which ran deep right up to the shore.

Both young men loved this sort of boat-camping. It was very like the trips FitzRoy had made with the whaleboats through Otway and Skyring Waters in the first year of his command aboard the *Beagle*, when he had slept on beaches beneath his sea cloak, finding it frozen hard over him in the mornings, and himself exhilarated by the experience.

For Darwin, the rough and truly dangerous conditions greatly appealed to the physically rugged side of his nature that had found its outlet at home in riding and shooting; it was a magnificent enlargement on the geologizing ramble he had made across Wales with Professor Sedgwick. Here, in addition to the cornucopia of natural phenomena to explore, there were wild savages to contend with, a sailing ship to call home, and weather severe enough to prompt him to grow a long beard whose tip he could see below his hand when he made a fist around it. It was the grandest adventure a boy ever had.

On successive nights they landed to make camp at deserted spots, the first at Shingle Point just west of the Murray Narrows, only to be quickly discovered and bothered by canoes full of aggressive Fuegians. These appeared unfamiliar with firearms, so the Englishmen's weapons were no deterrent. Rather than spend the night holding them at bay or in conflict, each time they packed up and moved on, trusting that the Fuegians would not follow them after dark. Nevertheless, even when they found campsites free of any sign of natives, they kept watch in turn through the night.

On his watch, sitting close to a fire in the dead dark of night, Darwin's imagination flamed. Fed by stories of the scuffles of the *Beagle*'s previous voyage, by tales of Cook and others, he was ever ready for an attack from wild savages.

It was my watch till one o'clock; there is something very solemn in such scenes; the consciousness rushes on the mind in how remote a corner of the globe you are then in . . . the quiet of the night is only interrupted by the heavy breathing of the men &

the cry of the night birds. — the occasional distant bark of a dog reminds one that the Fuegians may be prowling, close to the tents, ready for a fatal rush . . . their courage is like that of a wild beast, they would not think of their inferiority in number, but each individual would endeavour to dash your brains out with a stone, as a tiger would be certain under similar circumstances to tear you.

The day after recording such febrile imaginings, nature made a rush for Darwin. His courageous and instant response got his name stamped into geography for the first time.

That day, the two boats entered the northern arm of the Beagle Channel and the scenery changed dramatically. The land along the north shore of the channel now rose steeply to the high, permanently snow-covered, jagged southern cordillera of the Andes. Cottagey Kincaid country gave way to waterlogged Himalayan vistas. Waterfalls poured from the heights, glaciers tumbled into the Beagle Channel only a few hundred yards from the men's oars, surrounding the boats with small icebergs, or "growlers." The land turned tundra-like, scraped clean of all but the most stunted, wind-bent vegetation. The water in the channel, right up to the deep edge of the land and in the dense shadows of the glaciers, was a dark blue-black. Both FitzRoy and Darwin remarked on the "beryl blue" sepulchral glow of light refracting through the glacial ice.

At midday they stopped to cook a meal on a sandy point of land "immediately in front of a noble precipice of solid ice," wrote FitzRoy, "the cliffy face of a huge glacier, which seemed to cover the side of a mountain, and completely filled a valley several leagues in extent." It was not a good spot for lunch. As they gathered around their fire, the edge of the glacier right in front of them suddenly calved.

Down came the whole front of the icy cliff, and the sea surged up in a vast heap of foam. Reverberating echoes sounded in every direction, from the lofty mountains which hemmed us in; but our whole attention was immediately called to great rolling waves which came so rapidly that there was scarcely time for the most active of our party to run and seize the boats before they were tossed along the beach like empty calabashes. By the exertions of those who grappled them or seized their ropes, they were hauled up again out of reach of a second and third roller; and indeed we had good reason to rejoice that they were just saved in time.

Darwin was among the first to jump up and grab the boats from the sudden waves that would undoubtedly have carried them off, stranding the men ashore. His instant grasp of the situation, quicker than most of the seamen in the group, and his action in helping save the boats, prompted a special consideration from FitzRoy.

The following day, the 30th, we passed into a large expanse of water, which I named Darwin Sound—after my messmate, who so willingly encountered the discomfort and risk of a long cruise in a small loaded boat.

FitzRoy later named the high mountain above the Beagle Channel Mount Darwin. Many bestowals upon geography by early explorers have been lost or changed over time (for instance, a number of Captain Cook's names for the headlands and harbors of the New Zealand coastline have, in recent, more politically correct times, been replaced by their original Maori names). But Cerro Darwin still rises 7,975 feet high in the Cordillera Darwin above Seno Darwin on modern maps of Chilean Tierra del Fuego.

Several days later, the boats reached the open Pacific. In "miserable weather" they turned around, traveling back along the southwest arm of the Beagle Channel. From a whaleboat,

FitzRoy's observations and surveying of this western end of the Beagle Channel were quick and relatively rough, but accurate enough to provide the basis of charts still in use today.

On February 5, at Shingle Point, close to the Murray Narrows—where they had earlier been disturbed by a group of Fuegians and decided to move camp—they met some of this same group again. Now they appeared to be "in full dress," as FitzRoy put it: covered with red and white paint, goose down and feathers. They also wore ribbons and scraps of red cloth—gifts from the *Beagle's* seamen handed out at Woollya if they hadn't been stolen since—but one of their women, "noticed by several among us as being far from ill-looking," was wearing one of Fuegia's dresses. "There was also an air of almost defiance among these people, which looked as if they knew that harm had been done." FitzRoy immediately grew anxious about settlers at Woollya.

They hurried on, rowing until it was too dark to see. They were in the boats again at daybreak. Sluicing through the Murray Narrows they saw more natives "ornamented with strips of tartan cloth or white linen, which we well knew were obtained from our poor friends. No questions were asked; we thought our progress slow, though wind and tide favoured us: but hurrying on, at noon we reached Woollya."

The beach was thick with canoes; a hundred Fuegians were gathered along the shore and wandering around the new settlement. "All were much painted, and ornamented with rags of English clothing, which we concluded to be the last remnants of our friends' stock."

FitzRoy's darkest visions seemed about to be realized when the two whaleboats touched the shore and the Fuegians ran down to the water and surrounded their crews, leaping and shouting at them.

But then Matthews appeared, his clothes and person intact, followed by York and Jemmy, still dressed in their English duds and looking as well as usual.

Fuegia Basket did not appear. She was in York's wigwam, they said. FitzRoy gave no further explanation, but the reason for her indisposition is obvious: her first nights inside the wigwam with York Minster had also been her first alone with him. He had almost certainly had sex with his twelve-year-old bride. He had probably raped her, repeatedly.

FitzRoy pulled Matthews into a whaleboat and instructed the crew to row them a little distance offshore. There, away from the noise and interruption, he heard his story, while the natives squatted on their haunches along the beach watching them, "reminding me," wrote FitzRoy, "of a pack of hounds waiting for a fox to be unearthed."

The young missionary's zeal and stoicism had collapsed in the face of the rude attention directed at him.

> Matthews gave a bad account of the prospect which he saw before him, and told me, that he did not think himself safe among such a set of utter savages as he found them to be.

At first there had been only "a few quiet natives." But three days after the English boats had left, canoes full of rowdier Fuegians had turned up to bedevil him night and day. They stole anything he left lying around his wigwam; others crowded inside it all day, chattering and imploring him to give them everything they saw. Some of them became belligerent when he refused. More than one, he told FitzRoy, "went out in a rage, and returned immediately with a large stone" implying that he would kill Matthews if he didn't hand over what had been demanded. Others amused themselves by teasing him, "making mouths at him," and holding him down and pulling the hair out of his face. For a few days Matthews had been able to leave Jemmy guarding his wigwam while he visited the women in theirs, where they fed him and asked nothing in return, but soon the besieging, importuning, thieving mob made it impossible for him to leave his wigwam. Despite

Jemmy's constant assurances to the contrary, Matthews told FitzRoy that he believed he would soon be killed.

York Minster, a superior physical specimen among the Fuegians, had not been bothered and had lost nothing. Fuegia had his protection against everything outside their wigwam. But Jemmy had been "sadly plundered, even by his own family." "My people very bad," he told FitzRoy mournfully, "great fool, know nothing at all, very great fool."

The seamen's prize garden had been wantonly trampled, despite Jemmy's attempt to explain its purpose, and—or perhaps because of—his best efforts to keep people off it.

FitzRoy decided to take Matthews away with him, and the formerly zealous missionary gave no argument. The seamen quickly retrieved his chest and cache of personal possessions from the cellar in his wigwam and got it all into the whaleboats. FitzRoy himself distributed some of Matthews' stores—axes, saws, knives, and nails—among the surrounding natives, as a measure of goodwill, hoping some of this might reflect on his three protégés who were to remain: Jemmy, York, and Fuegia.

FitzRoy promised Jemmy and York he would return in a few days to see how they were getting on. Then the Englishmen pushed off in the whaleboats.

> When fairly out of sight of Woollya, sailing with a fair wind towards the *Beagle*, Matthews must have felt almost like a man reprieved, excepting that he enjoyed the feelings always sure to reward those who try to do their duty, in addition to those excited by a sudden certainty of his life being out of jeopardy.

Matthews had spent more than a year and come a long, hard way, with the blessing and endowments of many patrons, to establish the mission settlement in Tierra del Fuego. He may indeed have felt a manic surge of relief at being rescued from his worst imaginings. But if he felt as happy as FitzRoy suggests, he was fulfilling all Darwin's doubts about him. "He is of an eccentric character & does not appear . . . to possess much energy," Darwin had confided to his diary days earlier, before they had left Matthews in his wigwam. "I think it very doubtful how far he is qualified for so arduous an undertaking."

FitzRoy does not record his own feelings at seeing Matthews so readily abandon his post and this crucial element of his noble design, so long hoped-for, fail so quickly and completely. His remarks about Matthews are charitable. Behind them his own bitter disappointment shouts in mute, raging relief.

A week later, after briefly surveying sections of the Wollaston and Navarin Islands and Ponsonby Sound, FitzRoy returned with a single whaleboat crew to Woollya.

The scene that met him allayed his "considerable anxiety." The whaleboat passed women fishing peacefully from canoes. On the beach at Woollya, FitzRoy met Jemmy, York, and Fuegia, dressed as usual, Fuegia even "clean and tidily dressed." There were only a few other natives present, and they seemed "quiet and well disposed." FitzRoy inspected the three wigwams and found them unchanged and in good repair. The formerly trampled garden had recovered, and was now even sprouting a few vegetables. Beside York and Fuegia's wigwam lay a partly completed canoe that York was building from planks left behind by the crew. The place looked peaceful and productive.

Jemmy complained (he did a lot of this) that many of his things had been stolen. But apart from a few lapses by his own brothers, the thieves had mostly been strangers, now gone, and he and his family were on good terms. Jemmy's mother came down to the boat to greet her son's benefactor—and her own, for she was wearing a dress Jemmy had brought for her.

When FitzRoy and his boat crew shoved off, he was hopeful. "I left the place, with rather sanguine hopes of their effecting among their countrymen some change for the better."

FitzRoy had done his best—his considerable utmost, with the full backing of the Royal Navy, his king, queen, and the blessings of many like-minded Christians of England. Surely, his faith in all he held to be true assured him, God would now take this seed and make it grow.

Darwin was not so easily persuaded.

It was quite melancholy leaving our Fuegians amongst their barbarous countrymen: there was one comfort; they appeared to have no personal fears. — But in contradiction to what has often been stated, 3 years has been sufficient to change savages, into, as far as habits go, complete & voluntary Europæans. — [but] I am afraid whatever other ends their excursion to England produces, it will not be conducive to their happiness. — They have far too much sense not to see the vast superiority of civilized over uncivilized habits; & yet I am afraid to the latter they must return.

Distracted as he had been from his survey work (one gets the feeling, reading FitzRoy's narrative, that he was never again able to bring to it the singlemindedness that had characterized his efforts before the whaleboat had been stolen, just over three years earlier), FitzRoy felt he had covered enough of eastern Tierra del Fuego. On February 26, 1833, the *Beagle* sailed out into a gale in the Strait of Le Maire and ran in heavy seas toward the Falkland Islands.

17

A year and more into this second voyage aboard the *Beagle*, FitzRoy brooded.

He had spent months distracted from his primary, ostensible purpose—surveying and mapping the southern South American coastline, east and west—with the settling of the Fuegians. That now done, unsatisfactorily, the work still to be accomplished loomed before him.

In addition to surveying the Falkland Islands, there were still gaps in his surveys of the Atlantic coast, uncertainties of longitude to be resolved—work he had put aside to take the Fuegians south during the summer season. He would have to sail north again and spend another year crawling along the *pampero* coast of Argentine Patagonia: thousands of miles that included vast river estuaries, gateways to the continent's developing interior, all of it potentially important for future trade and safe navigation, yet it was impossible to sail every mile of it. There was still the whole world to circumnavigate: all those chronometers bought and daily tended to mark off an unbroken chain of longitudes around the globe. The entire voyage was supposed to last just three years and with almost half that time gone, he was still mired on the Atlantic coast. The job seemed endless. He had brought it on himself, it was what he had wanted, an important peacetime commission. Now it seemed impossible.

He was unfairly overburdened, FitzRoy felt. On the first voyage, the survey area—just Tierra del Fuego—was far smaller, and the *Beagle* had been one of three vessels, with the *Adventure* and the *Adelaide*, carrying out the work. Now, with a single ship, he began to despair of being able to complete the work and justifying the time and expenditure of this voyage as more than simply a repatriation effort on behalf of his three Fuegians.

The young aristocrat's attenuated sensibility stretched and tightened. The captain's cabin aboard the *Beagle* became once again the fraught isolation chamber inside which Pringle Stokes had succumbed to despair.

Within days of arriving at the Falklands, an accident occurred that further isolated him. There was a standing order aboard the *Beagle* that no one venture far ashore alone. FitzRoy's clerk, Edward Hellyer, went off to shoot in the company of a Frenchman from a wrecked French whaler, *Le Magellan*, whose crew were living in tents ashore, awaiting a ride to the South American mainland. After the two men had walked a mile alongshore together, the Frenchman returned to his camp and Hellyer went on alone. An hour later, clothes, a gun, and a watch were found near the shore. Seamen from the *Beagle* and some of the French crew ran along the beach looking for Hellyer, others pulled alongshore and through the kelp beds in whaleboats. After a long search, his body was found, naked, entangled in kelp. Bynoe, the *Beagle*'s surgeon, and his French counterpart, tried everything to revive the boy. A duck was found dead in the kelp near Hellyer's body; his gun lying on the beach had been discharged.

FitzRoy expressed his sorrow in the most conventional terms when he came to write up his narrative years later: "Mr Hellyer had been much with me, both as my clerk and because I liked his company, being a gentlemanly, sensible young man." But Hellyer's death was a significant loss to him. The complex job of captain's clerk was a plum one, not unlike the position of personal assistant to a CEO. It generally went to young men who could combine an impressive array of secretarial talents with an unusually personable, sensitive nature. Finding such help among a ship's company was difficult, and FitzRoy lost Hellyer at a time when his task threatened to overwhelm him. A deeply emotional man, he was made unhappier by the belief that the boy had shot the duck, of a species they had not seen before, for his captain, and, when it had fallen into the water, stripped and swum after it.

The boy was buried, after a melancholy procession, on a lonely and dreary headland. FitzRoy's burden only felt heavier.

Days later, an opportunity that seemed tailor-made to improve his predicament presented itself. An American sealing schooner, the 170-ton *Unicorn*, had put into the Falklands after a six-month cruise in Cape Horn waters with her hold empty, her master and part owner ruined. An unusually stormy summer season (the great gale of January 13 that had knocked the *Beagle* down had driven *Le Magellan* ashore in the Falklands) had kept the *Unicorn* bottled up in harbors riding out gales for sixty-seven days, preventing the taking of any seals, and her master was ready to sell. The hire of the two schooners, the *Liebre* and the *Paz*, to cover some of the coast to be surveyed had worked so well the year before that FitzRoy's desire to purchase the *Unicorn* was "unconquerable." The *Beagle*'s carpenter, Jonathan May, looked over the ship and reported back that she was sound, her construction first-class: fastened with copper spikes, planked of good oak. She was roomy, easily handled by a small crew, and a good sea boat. She was more: pretty as a yacht, she was a seductress as ships go. FitzRoy seems to have fallen in love with her, in the way that seamen can find their hearts stolen by two or three ships that they will remember for a lifetime, vessels that somehow contrive to be more than their wood and metal and canvas. She was beautiful, and he needed her.

He bought the *Unicorn* for £1300 out of his own pocket. He

hoped that his purpose in doing so—to aid the *Beagle* in covering the territory to be surveyed, to speed up the work, enabling the cruise to continue on around the world without years of delay— would meet with Admiralty approval and see him reimbursed. He renamed her *Adventure* in honor of the *Beagle*'s former consort.

It seemed the right thing to do at the time.

While FitzRoy and his crew rowed around the islands surveying from the whaleboats, Darwin explored ashore, often in the company of Syms Covington. He found the Falkland Islands dull and uninteresting: "We have never before stayed so long at a place & with so little for the Journal."

He spent six days riding across East Falkland with gauchos who had come to the islands to catch some of the wild cattle (descendents of shipborne livestock) that roamed there. "The inhabitants are a curious mixed race," Darwin wrote of the motley population of shipwrecked sailors, whalers, and sealers of many nations, gauchos, Argentine and Spanish colonists, three women, "two of them negresses," and the single resident Englishman, Mr. Dixon, under whose flag the islands were in temporary possession. "Their habitations are in a miserable condition & deficient in almost every accomodation. The place bespeaks what it has been, viz a bone of contention between different nations."

Although Darwin's early focus was geology, the jottings in his notebook while in the Falkland Islands show that he was beginning to be irresistibly drawn into the mystery of the differences between species and their geographic distribution.

March 2: To what animals did the dung beetles in South America belong—Is not the closer connection of insects and plants as well as this fact point out closer connection than Migration.

Tuesday 12th: Horses fond of catching cattle—aberration of instinct.

21st: Saw a cormorant catch a fish & let it go 8 times successively like a cat does a mouse or otter a fish.

22nd: Migration of (Upland) Geese in Falkland Islands as connected with Rio Negro?

And in his published *Journal of Researches* (later renamed *The Voyage of the Beagle*), Darwin wrote:

> The only quadruped native to the island is a large wolf-like fox which is common to both East and West Falkland. I have no doubt it is a peculiar species, and confined to this archipelago; because many sealers, gauchos, and Indians, who have visited these islands, all maintain that no such animal is found in any part of South America. . . . As far as I am aware, there is no other instance in any part of the world of so small a mass of broken land, distant from a continent, possessing so large an aboriginal quadruped peculiar to itself. Their numbers have rapidly decreased. . . . Within a very few years after these islands shall have become regularly settled in all probability this fox will be classed with the dodo, as an animal which has perished from the face of the earth.*

FitzRoy and Darwin disagreed about the foxes. "Naturalists," wrote FitzRoy, referring to the only naturalist whose opinion he had sounded on the subject, "say that these foxes are peculiar to this archipelago, and they find difficulty accounting for their presence in that quarter only." FitzRoy thought them very similar, different only in the shading of their coats, to the Patagonian foxes he had seen. He believed foxes from the mainland had been carried to the islands aboard large chunks of ice riding the current that sets between southern Tierra del Fuego and the Falklands. The Falkland foxes were simply variants of the same species, he reasoned.

---

*The fox still roams the Falklands.

The animals they saw in the Falkland Islands (and everywhere else, notably the Galapagos Islands), and their variation from the genus species they were familiar with, made for frequent discussions between the two scientist messmates at the captain's table aboard the *Beagle*.

FitzRoy generally disagreed with Darwin's early musings about the way animals might change from one place to another:

Rats and mice were probably taken to the Falklands by the earlier navigators who landed there, whose ships were often plagued with their numbers. That they have varied from the original stock in sharpness of nose, length of tail, colour, or size, is to be expected . . . but to fancy that every kind of mouse which differs externally from the mouse of another country is a distinct species, is to me as difficult to believe as that every variety of dog and every variety of the human race constitute a distinct species. I think that naturalists who assert the contrary are bound to examine the comparative anatomy of all these varieties more fully, and to tell us how far they differ. My own opinion is, judging from what I have gathered on the subject from various sources, that their anatomical arrangement is as uniformly similar as that of the dogs and the varieties of man.

However much he and Darwin disagreed in their thinking, this sort of engaging debate is what FitzRoy had longed for on the *Beagle*'s first voyage, what he had appealed to Beaufort to supply him with on this one. It brought him out of himself and kept him connected to the world.

But for Darwin, these constant discussions, turning over the findings and observations of a voyage around the world with a keen scientific mind that often sparked and challenged him with an opposing view, were of incalculable benefit.

Early in April, the *Beagle* and the *Adventure* sailed for Rio Negro on the Argentine mainland.

*   *   *

Ten months later, the *Beagle* was back in Tierra del Fuego. FitzRoy had further chronometric readings to make at Port Famine inside the Strait of Magellan, and he still needed to survey the Fuegian coast below the eastern entrance to the strait— the area around the rolly open anchorage off Cape Santa Iñez that the *Beagle* had fled in foul weather in December 1832.

For many of those intervening ten months, Darwin traveled widely inland. One of these excursions, overland from Rio Negro to Buenos Aires, was an exciting and dangerous adventure. The Argentine General Rosas was engaged in "a bloody war of extermination against the Indians," but "so fine an opportunity for geology was not to be neglected," Darwin wrote his sisters. In company with an English trader, James Harris, a peon guide, and a band of gauchos, he traveled between *postas*, Spanish army camps; Darwin and Harris roamed deeper into the country than any previous European travelers. The gauchos, a species unto themselves, made a vivid impression on the young Englishman.

> They are generally tall and handsome, but with a proud and dissolute expression of countenance. They frequently wear their moustaches, and long black hair curling down their backs. With their brightly coloured garments, great spurs clanking about their heels, and knives stuck as daggers (and often so used) at their waists, they look a very different race of men from what might be expected from their name of Gauchos, or simple countrymen. Their politeness is excessive: they never drink their spirits without expecting you to taste it; but whilst making their exceedingly graceful bow, they seem quite as ready, if occasion offered, to cut your throat.

Darwin rejoined the *Beagle* in Montevideo, where, on December 5, 1833, he "took a farewell of the shore & went on board."

On February 2, 1834, the ship anchored in Port Famine. The harbor of Pringle Stokes's suicide (now 5½ years past) was still an unrelievedly depressing place. "I never saw a more cheer-less prospect," Darwin later wrote of this place, where it seemed to rain continuously, and ashore he found a "death-like scene of desolation [that] exceeds all description . . . everything was dripping with water; even the very Fungi could not flourish."

For the next three weeks, FitzRoy worked the ship down the eastern Fuegian coast, through the Strait of Le Maire and back into very familiar waters: up Nassau Bay, through Goree Road, and into the Beagle Channel. Only the *Beagle*'s whaleboats, not the ship herself, had floated upon her namesake waters before now.

On February 25, headed shoreward in a whaleboat with a group of seamen, Darwin passed a canoe holding six "Yapoo Tekeenicas." More than ever, he was struck by the existence of unreconstructed Fuegians in their natural environment. His experiences, his thinking and discussions with FitzRoy over the past year, led him now to a deeper reflection on the extreme low-liness of their condition. They seemed to exhibit a disregard for the meagerest of niceties that might lift a human one notch above the animal state. That day he wrote in his diary:

I never saw more miserable creatures; stunted in their growth, their hideous faces bedaubed with white paint & quite naked. — One full aged woman absolutely so, the rain & spray were dripping from her body; their red skins filthy & greasy, their hair entangled, their voices discordant, their gesticulation violent & without any dignity. Viewing such men, one can hardly make oneself believe that they are fellow creatures placed in the same world. . . . Although essentially the same creature, how little must the mind of one of these beings resemble that of an educated man. What a scale of improvement is comprehended between the faculties of a Fuegian savage & a Sir Isaac Newton — Whence have these people come? Have they remained in the same state since the creation of the world? What could have tempted a tribe of men leaving the fine regions of the North to travel down the Cordilleras the backbone of America, to invent & build canoes, & then to enter upon one of the most inhospitable countries in the world. . . . Nature, by making habit omnipotent, has fitted the Fuegian to the climate & productions of his country.

Just as they exchanged views about Falkland foxes, Darwin and FitzRoy surely had countless dinner table discussions along exactly these lines. Both men found the Fuegians equally instructive grist for later conclusions.

The only thing more fascinating than seeing these time-capsule humans in their natural habitat—like cave dwellers from a diorama come to life before one's eyes—was to dress them up as English gentlefolk, drop them off in the wild, and see how long the patina of civilization could last. The ship was now back in Jemmy Button's country, and FitzRoy—along with everyone else aboard—was eager to know what had become of the three Fuegians they had left ashore almost a year earlier. On March 5, after days of beating to windward through the Beagle Channel, the Beagle rode the tide through Murray Narrows and came to anchor at Woollya.

"Not a living soul was visible any where," wrote FitzRoy, noting only what he was looking for on land; the ship had been followed through the narrows by seven canoes full of Fuegians waving bows and arrows. Going ashore with some men he found the *Beagle*-built wigwams were still standing, undisturbed but empty, and showing no signs of recent habitation. The garden appeared trampled and neglected, but the seamen dug up some turnips and potatoes "of moderate size" which the captain later ate for dinner.

Back aboard the ship, FitzRoy feared the worst, particularly in view of the aggressive behavior of the natives in the canoes. But an hour or two later, three more canoes were sighted, paddling strongly for the ship from a nearby island. FitzRoy raised a telescope in their direction.

I saw that two of the natives in them were washing their faces, while the rest were paddling with might and main: I was then sure that some of our acquaintances were there, and in a few minutes recognized Tommy Button, Jemmy's brother. In the other canoe was a face which I knew yet could not name. "It must be some one I have seen before," said I, when his sharp eye detected me, and a sudden movement of the hand to his head (as a sailor touches his hat) at once told me it was indeed Jemmy Button—but how altered!

In shame, Jemmy kept his back to the ship until the canoe came alongside. Up he scrambled onto the deck. The fat, vain dandy was no more. In his place, observed Darwin, was a thin, haggard savage, with long matted hair, and naked, except for a skin around his waist. "I could hardly restrain my feelings," wrote FitzRoy, "and I was not, by any means, the only one so touched by his squalid, miserable appearance."

With an unfailing sense of what was most important, the Englishmen hurried Jemmy below to be clothed. In half an hour he was sitting at the captain's table. He ate lunch with his manners unimpaired, using his knife and fork as correctly as ever. FitzRoy thought he looked ill, but Jemmy assured him that he was "hearty, sir, never better." He hadn't been sick a day since he had last seen them, and ate "plenty fruits, plenty birdies, ten guanacos in snow time," and "too much fish."

What had happened to York and Fuegia? FitzRoy asked.

After the *Beagle* had left, the year before, Jemmy told them, other Fuegians—Jemmy's enemies the Oens-men, not of his country—hearing of the settlement, had raided the camp at Woollya. They looted whatever Jemmy and his family had not been able to escape with in their canoes. York had managed to save most of his belongings, including the large canoe FitzRoy had seen him building beside his wigwam. He and Fuegia then urged Jemmy and his family to move with them, with all their remaining belongings, to York's country, farther west, where he had first been taken from. They traveled as far as Devil Island at the western end of the Beagle Channel where they came upon York's brother and other members of his tribe, the Alacalufes (or the Alikhoolips, as FitzRoy called them). There, while Jemmy and his family slept on Devil Island, York made off with all his worldly goods, leaving him in his naked, original state. An act of consummate villainy, Darwin thought. FitzRoy, when he heard this, saw in it considerable cunning.

York's fine canoe was evidently not built for transporting himself alone; neither was the meeting with his brother accidental. I am now quite sure that from the time of his changing his mind [the year before, in January 1833, when the *Beagle* had spent weeks trying to fight its way west below Cape Horn toward York's country around March Harbour], and desiring to be placed at Woollya, with Matthews and Jemmy, he meditated taking a good opportunity of possessing himself of every thing; and that he thought, if he were left in his own country without Matthews, he would not have many things given to him, neither would he know where he might afterwards look for and plunder poor Jemmy.

They must have asked Jemmy if he wanted to come away with them again, because both FitzRoy and Darwin wrote that he was happy and contented with his life and had no wish to change it or to return to England.

After lunch Jemmy visited with members of the crew. He had brought two otter skins, one for FitzRoy, and the other for the captain's steady coxswain, James Bennet, who had overseen all the Fuegians' arrangements in England and spent so much time with them there.

While Jemmy's English appeared as good as ever, he told FitzRoy that his Fuegian was still poor, but that he spoke with his family now in both languages, and they appeared to understand him. They were speaking a little English. This seemed like the thin end of the wedge that FitzRoy had always hoped for.

In the evening, another canoe came alongside the ship with an attractive young woman in it. She was crying fearfully, unstoppably, for Jemmy Button. It was his wife, Jemmy told the crew. Immediately she was showered with gifts—shawls, handkerchiefs, and a gold-laced cap—but would not stop crying until Jemmy appeared on deck close by. His brother, Tommy, also felt the visit had gone on long enough, for he began to bellow in his stentorian Fuenglais: "Jemmy Button, canoe, come!"

Jemmy paddled away with his family for the night. They did not head for Woollya. The wigwams there had been too good: they were too high and cold in the winter. And the terrain around the settlement—a parklike setting that appealed to English tastes, sketched and painted by the *Beagle*'s artists—made them too vulnerable to attack. They steered instead for nearby "Button Island," as all now called it, where Jemmy felt better off.

After farewells and more present giving the next day, the *Beagle* sailed away. Jemmy sailed with it for a short distance, until his wife's violent crying got him back into her canoe. He had developed a fondness for ships and the shipboard life. He would not forget it.

Darwin was glad to see the last of the Fuegians. Scientifically they fascinated him, but he had grown sick of their incessant, importuning "yammerschoonering." "Saying their favorite word in as many intonations as possible, they would then use it in a neuter sense, and vacantly repeat, 'yammerschooner.' On leaving some place we have said to each other, 'Thank Heaven, we have at last fairly left these wretches!' when one more faint halloo from an all-powerful voice, heard at prodigious distance, would reach our ears, and clearly could we distinguish—'*Yammerschooner.*'"

To Darwin, FitzRoy's great experiment seemed to have failed utterly. Prefiguring his long cogitation on adaptation and the pecking order of evolving species, the young naturalist intuited a crucial impediment to the improvement of the aborigines of Tierra del Fuego:

> The perfect equality of all the inhabitants will for many years prevent their civilization: even a shirt or other article of clothing [given to one as a gift] is immediately torn into pieces [to be shared].—Until some chief rises, who by his power might be able to keep to himself such presents as animals &c &c, there must be an end to all hopes of bettering their condition.

FitzRoy would not admit it. He persisted, even now, in seeing a hopeful outcome.

> I cannot help still hoping that some benefit, however slight, may result from the intercourse of these people, Jemmy, York, and Fuegia, with other natives of Tierra del Fuego. Perhaps a ship-wrecked seaman may hereafter receive help and kind treatment from Jemmy Button's children; prompted, as they can hardly fail to be, by the traditions they will have heard of men of other lands; and by an idea, however faint, of their duty to God as well as their neighbour.

Jemmy Button would prove how terribly wrong FitzRoy was.

18

FitzRoy never counted on Admiralty support for his extracurricular hiring and buying of additional vessels to aid his surveying efforts, but he had certainly hoped for it. Financially he needed it.

"I believe that their Lordships will approve of what I have done," FitzRoy had written to Captain Beaufort, "but if I am wrong, no inconvenience will result to the public service, since I alone am responsible for the agreement . . . and am able and willing to pay the stipulated sum."

The dispatch secretary at the Admiralty underlined the phrase "am able and willing" and asked Beaufort for a report on this irregularity at the next Board Day meeting of the Lords. Beaufort's response was as supportive of his man out mapping the edge of the world as he could make it.

There is no expression in the Sailing Orders, or surveying instructions, given to Commander FitzRoy which convey to him any authority for hiring and employing any vessels whatever.

On the other hand, there can be no doubt that by the aid of small craft he will be sooner and better able to accomplish the great length of coast which he has to examine — and which seems to contain so many unknown and valuable harbours; —

especially if he finds it necessary to trace the course of a great river, which had been reported to him as being navigable almost to the other side of America.

It may also be stated to their Lordships that the *Beagle* is the only surveying ship to which a smaller vessel or Tender has not been attached.

In the last paragraph, Beaufort invokes the ideal and more usual practice of all exploring voyages since Man first set out on logs across a lagoon: sailing in company with another vessel as insurance. Columbus did it, as did Magellan and Cook. It was unusual and risky to send a ship alone to explore a remote corner of the world. If she struck a reef, her crew could easily perish unless a companion ship stood nearby. Of Magellan's five-vessel squadron that set out from Spain to circumnavigate the world in 1519, only one ship, battered and manned by 31 of the 270 men who had started the voyage, returned to Seville. On her first voyage to Tierra del Fuego, the *Beagle* had sailed with HMS *Adventure* commanded by Captain King, and this, as Beaufort points out, was standard practice with surveying missions.

In another letter, FitzRoy reminded Beaufort of the time slipping by, and the opportunity made possible by a sister vessel.

I cannot help feeling rather strongly that the *Adventure* and *Beagle* have been several years about this survey, and that Foreigners as well as Englishmen are anxiously expecting the results of the "English Survey." Officers acquainted with these countries are now employed who may be elsewhere in a short time. Chronometers will not continue to go well for *many* years, without cleaning—The *Beagle* has many measurements to make and much work to do in the Pacific. And a certain troubled spirit and conscience is always goading me to do all I can, for the sake of doing what is *right*; without seeking for credit, or being cast down because everyone does not see things in the same light. These are some of the reasons which occasion my outgoings.

What is *now* left undone, will long be neglected. Not only the character of those actually engaged in the survey will suffer, but the credit of the English as surveyors will be injured.

Despite his avowed willingness to bear the expense, FitzRoy was banking on Admiralty approval. After purchasing the *Unicorn* and rechristening her *Adventure*, he bought stores and gear to refit her from the wrecked French whaler *Le Magellan*, had her careened (hauled over on her side at the water's edge, exposing the bottom of the hull) and her bottom coppered, and fitted her out with no expense spared to sail in convoy with the *Beagle* around the world. He was hugely excited by the opportunities made possible with a second ship. She would virtually double his charting and exploring of the Pacific islands; she would offer his crew safety; and she might help to make him famous. FitzRoy knew his history, and he saw the *Beagle* and her consort *Adventure* taking their place in the pantheon of landmark voyages. The supplies aboard both vessels, and the mandate of his mission, gave FitzRoy the chance to make an exceptional mark on the world. He believed that no one on Earth at that moment had the chance and the wherewithal to open up the globe, to delve into its natural and scientific mysteries, as he now hoped to do. All these hopes lay gathered in his purchase of the schooner.

He was inordinately proud and protective of his new little ship. As the *Beagle* and the *Adventure* neared the western shores of Tierra del Fuego, ready to sail out into the Pacific, poor weather and visibility kept them pinned inside the Furies, a rock-studded constellation of small islets that posed a death trap for ships. Night came on with heavy, view-obliterating rain squalls. There was one safe anchorage in the area, a tiny cove with room in it for a single vessel. FitzRoy sent the *Adventure* in to shelter in safety while he kept the *Beagle* underway all night, tacking and wearing back and forth through the black foul weather and racing tides in a space of four square miles. It was a purely emotional, and strikingly unsound and unseaman-like decision to keep the mother ship—the bluffer, unhandier, less efficient sailer, with the greater amount of stores and number of men aboard—turning and turning about between the rocks through the long black hours. But he carried it off with his usual consummate, relished seamanship:

> It was necessary to keep under a reasonable press of sail part of the time, to hold our ground against the lee tide; but with the ebb we had often to bear up and run to leeward, when we got too near the islets westward of us. In a case of this kind a ship is so much more manageable while going through the water than she is while hove-to, and those on board are in general so much more on the alert than when the vessel herself seems half asleep, that I have always been an advocate for short tacks under manageable sail, so as to keep as much as possible near the same place, in preference to heaving-to and drifting.
>
> When the day at last broke . . . we saw the Adventure coming out to us from the cove where she had passed the night, and then both vessels sailed out of the Channel, past Mount Skyring and all the Furies, as fast as sails could urge them. At sunset we were near the Tower Rocks, and with a fresh north-west wind stood out into the Pacific, with every inch of canvas set which we could carry.

The Furies have always made strong men quail. Sixty-two years later, in March 1896, Joshua Slocum, the first man to sail alone around the world, found himself trapped among them in his 37-foot sloop, at night in a roaring gale.

> Night closed in before the sloop reached land, leaving her feeling the way in pitchy darkness. I saw breakers ahead before long. At this I wore ship and stood offshore, but was immediately startled by the tremendous roaring of breakers again ahead and on the lee bow. This puzzled me, for there should have been no broken water where I supposed myself to be. I kept off a good bit, then wore round, but finding broken water also there,

threw her head again offshore. In this way, among dangers, I spent the rest of the night. Hail and sleet in the fierce squalls cut my flesh till the blood trickled over my face; but what of that? It was daylight, and the sloop was in the midst of the Milky Way of the sea, which is northwest of Cape Horn, and it was the white breakers of a huge sea over sunken rocks which had threatened to engulf her through the night. It was Fury Island I had sighted and steered for, and what a panorama was before me now and all around! It was not the time to complain of a broken skin. What could I do but fill away among the breakers and find a channel between them, now that it was day? Since she had escaped the rocks through the night, surely she would find her way by daylight. This was the greatest sea adventure of my life. God knows how my vessel escaped. . . . The great naturalist Darwin looked over this seascape from the deck of the *Beagle*, and wrote in his journal "Any landsman seeing the Milky Way would have nightmares for a week." He might have added, "or seaman" as well.

Darwin did indeed find the Furies and their environs nightmarish; this is what he actually wrote.

Outside the main islands, there are numberless rocks & breakers on which the long swell of the open Pacific incessantly rages. — We passed out between the "East & West Furies"; a little further to the North, the Captain from the number of breakers called the sea the "Milky Way". — The sight of such a coast is enough to make a landsman dream for a week about death, peril, & shipwreck.

So it was FitzRoy who, that night, coined the name by which seamen ever afterward called this deadly scattering of rocks.

The frightening view astern was Darwin's, FitzRoy's, and the rest of both ships' crews' last sight of Tierra del Fuego.

The *Adventure* proved a fast sailer, and even Darwin caught some of FitzRoy's pride and enjoyment in their consort: "The

Adventure kept ahead of us, which rejoiced us all, as there were strong fears about her sailing. It is a great amusement having a companion to gaze at."

But he was aware of the price being paid, and it worried him. "He [FitzRoy] is eating an enormous hole into his capital for the sake of advancing all the objects of the voyage," Darwin wrote home. "The schooner which will so very mainly be conducive to our safety he entirely pays for."

So FitzRoy was emotionally unprepared for the Admiralty's response, which finally reached him on the west coast of Chile.

> Their Lordships do not approve of hiring [and buying] vessels for the service and therefore desire that they may be discharged as soon as possible.

There would be no reimbursement.

By the time he let go the other two schooners, the *Pax* and the *Liebre*, which he had hired the year before to help him survey the South American Atlantic coast, FitzRoy had spent a total of £1680 to charter them, and more to refit them for the work. At a time when the average per capita income in Britain was about £20 per year, this was a fortune, lost in the zealous service of his commission. Now he had spent almost as much for the *Adventure* alone, and more to outfit her. He was so reluctant to give her up, and "all my cherished hopes," that he held onto her a little longer, hoping for an official change of heart. It did not come.

Darwin believed the "cold manner" of the Admiralty's response to its lone captain was "solely . . . because he is a Tory." Maybe. The Whigs, Britain's liberal reforming party, had recently been elected to office after a long hiatus from power, and the mood in government was set against the more conservative Tories, with whom FitzRoy's aristocratic family was traditionally associated. But more likely he was feeling the Admiralty's indifference or even antipathy to his mission, which, after all, had been engineered by his influential uncle. The Lords had given him a ship, outfitted it handsomely, and let him go; that was enough. Beaufort seems to have been quite alone in his strong championing of FitzRoy and his voyage. And certainly nobody gave any thought to opportunities afforded the *Beagle*'s naturalist, an unknown student, the captain's supernumerary indulgence.

Finally, the cost of buying and running the *Adventure* began to exceed FitzRoy's income and drive him into debt, and "after a most painful struggle," he found a buyer in Chile and sold it. Through being "dispirited and careless" he mismanaged the sale, getting close to what he had paid for the ship, but far less than he'd spent on her outfitting and renovation. It was a double loss that depressed him deeply.

The *Beagle* spent the winter months refitting in Concepción and Valparaiso. At times FitzRoy and his officers moved ashore to collate their surveys and draw their charts in good light and peace and quiet, away from the bustle aboard ship. He tried to bury himself and his disappointment in the work, but he was besieged by Chilean hospitality: constant invitations to entertain the captain and his officers distracted him and brought with them the obligations to return such favors in kind. His mood continued to tumble. He snapped viciously at Darwin, causing the second of their two quarrels, which Darwin remembered years later.

At Conception in Chile, poor FitzRoy was sadly overworked and in very low spirits; he complained bitterly to me that he must give a great party to all the inhabitants of the place. I remonstrated and said that I could see no such necessity on his part under the circumstances. He then burst out in a fury, declaring that I was the sort of man who would receive any favours and make no return. I got up and left the cabin without saying a word, and returned to Conception where I was then lodging. After a few days I came back to the ship and was received by the Captain as cordially as ever.

During that winter, FitzRoy's depression reached a state that seemed to Darwin to be "bordering on insanity." He refused to visit or be visited. He stopped eating. Darwin described his condition in a letter home: "a morbid depression of spirits, & a loss of all decision & resolution. The Captain was afraid that his mind was becoming deranged."

FitzRoy, the consummate navigator, had lost his way. He saw gaps in his information, gaps in his charts of the coasts, and became convinced he would have to sail south again and spend another season—their fourth—in Tierra del Fuego, a prospect that all aboard, including FitzRoy, dreaded. He lost sight of where to draw the line. He was succumbing to the overload that had driven Pringle Stokes to despair and suicide. He was having a nervous breakdown.

Bynoe, the ship's surgeon, told him he was suffering from overwork, that he should ease up and he would recover. But FitzRoy would not believe him. He felt madness burrowing its way through him. He spoke of the suicide of his uncle who had slashed his own throat, and he grew convinced that a hereditary susceptibility to that same maggoty decay was now attacking his mind. Both he and Bynoe were right: overwork, strain, uncertainty, and disappointment had crushed his delicate mental foundation and sent him spinning into a void.

He finally did the unthinkable: he relieved himself of command, appointing First Lieutenant Wickham commander of the *Beagle*. FitzRoy's instructions from the Admiralty always allowed for such an occurrence and were eerily clear about what was then to happen:

In the event of any unfortunate accident happening to yourself, the officer on whom the command of the Beagle may in consequence devolve, is hereby required and directed to complete, as far as in him lies, that part of the survey on which the vessel may be then engaged, but not to proceed to a new step in the voyage; as, for instance, if at that time carrying on the coast survey on the western side of South America, he is not to cross the Pacific, but to return to England by Rio de Janeiro and the Atlantic.

There would, in this case, be no continuing the voyage around the world.

Darwin wrote home of his profound disappointment, and his conflicted feelings about the voyage's growing length.

> As soon as the captain invalided, I was at once determined to leave the *Beagle*; but it was quite absurd what a revolution in five minutes was effected in all my feelings. I have long been grieved and most sorry at the interminable length of the voyage (although I never would have quitted it). But the minute it was all over, I could not make up my mind to return—I could not give up all the geological castles in the air which I had been building. . . . One whole night I tried to think over the pleasure of seeing Shrewsbury again, but the barren plains of Peru gained the day.

It was Lieutenant Wickham, FitzRoy's loyal second-in-command, who brought his captain round. He pointed out that the Admiralty's instructions for the survey of the southwest coast of South America were to do not all of it but as much as was conveniently possible within a reasonable period—and then to proceed across the Pacific. If he took command, Wickham said, nothing would induce him to spend more time in Tierra del Fuego. So what was to be gained by the captain's resignation? he asked FitzRoy. Wickham urged him to reconsider, to accept the sufficiency of what they had already done, to continue the mission by crossing the Pacific and returning to England at the conclusion of the circumnavigation all hoped to achieve. FitzRoy pondered this a short while and then agreed. He withdrew his resignation. Darwin wrote home with the good news.

To have endured Tierra del Fuego and not seen the Pacific would have been miserable. . . . When we are once at sea, I am sure the captain will be all right again. He has already regained his cool inflexible manner, which he had quite lost.

FitzRoy spent another year surveying the west coast of South America, but he stayed north of Tierra del Fuego, ranging between the temperate Patagonian island of Chiloé and the tropical waters above Lima, Peru.

On September 7, 1835, almost four years after leaving England, the *Beagle* at last sailed away from the continental coast of America, out onto the vast and storied Pacific.

History had wobbled for a moment as FitzRoy's despair got the better of him; the *Beagle* had almost turned around and sailed for home. But then her captain recovered, and she pointed her bow northwest toward a small scattering of islands on the equator, and history shifted its weight onto the *Beagle*'s unsuspecting natural philosopher.

19

Early voyagers called them the Enchanted Islands, but not because of anything they offered a passing sailor. They were hard to find, elusive, chimerical. They lay in a belt of fitful winds and humid cloud known as the Doldrums. Strong ocean currents played around them. With skies too overcast for celestial navigation, their ships drifting in unknown directions and spinning on ocean gyres, early navigators felt as if their instruments and the sea itself around these islands were bewitched. FitzRoy's task was to fix their location exactly.

Darwin's was to discover the clues that would underwrite his enduring fame. He would do it quickly: his momentous visit to the Galapagos Islands lasted just thirty-four days, and it would be more than a year later until he had any idea what he had really found there.

After the lushness of the tropics and the epic grandeur of Patagonia and Tierra del Fuego, Darwin found the islands disappointing. They were hot and uncharming. "All the plants have a wretched, weedy appearance, and I did not see one beautiful flower." The flowers he did find were "insignificant, ugly little flowers." He took great pains to collect insects but, "excepting Tierra del Fuego, I never saw in this respect so poor a country.

Even in the upper and damp region I procured very few," and those he found were "of very small size and dull colours."

The tortoises were more fun.

The inhabitants believe that these animals are absolutely deaf; certainly they do not overhear a person walking close behind them. I was always amused when overtaking one of these great monsters, as it was quietly pacing along, to see how suddenly, the instant I passed, it would draw in its head and legs, and uttering a deep hiss fall to the ground with a heavy sound, as if struck dead. I frequently got on their backs, and then giving a few raps on the hinder part of their shells, they would rise up and walk away; — but I found it very difficult to keep my balance.

He found the iguana lizards "most disgusting."

He noticed what everyone notices, or used to, in the Galapagos Islands: "The birds are strangers to man and think him as innocent as their countrymen the huge tortoises. Little birds, within three or four feet, quietly hopped about the bushes and were not frightened by stones being thrown at them. Mr King killed one with his hat and I pushed off a branch with the end of my gun a large hawk."

Most notably, Darwin observed and collected a number of birds with very different beaks. At the time, he believed these different beaks indicated different genera, or kinds, of birds: thrushes, finches, blackbirds.

It was not until much later, when his specimens had been examined by experts in England, that Darwin learned what was remarkable about his collections from the Galapagos Islands: the majority of all the animals and flowering plants were aboriginal. They were not found anywhere else.

While Darwin fossicked and collected, the *Beagle* sailed through the islands on its surveying mission.

One of the ship's anchorages was Post Office Bay on Floreana

Island. The name had sprung from a custom established by whaling ships, mostly those from Nantucket and New Bedford, that had been calling at the Galapagos Islands since the 1780s, almost fifty years before the *Beagle*'s arrival. Not only did the islands lie across the migratory path of whales; they provided the whaleships with large numbers of tortoises, which lumbered around their decks until slaughtered for fresh meat. In this sheltered bay, a whale-oil barrel had been erected with a small roof over it to hold letters which could be deposited and collected by newly arrived and homebound vessels. On voyages often lasting three or four years, such a mail drop was a treasure trove for seamen anxious for news of their families.

"Since the island has been peopled the box (barrel) has been empty, for letters are now left at the settlement," wrote FitzRoy. He was mistaken. The tradition held among the whalers for much of the nineteenth century and was continued by cruisers aboard sailing yachts well into the twentieth century. In December 1928, American William Albert Robinson, bound from New York around the world aboard the 32-foot ketch *Svaap*, posted a letter here. The barrel in which he left his letter was "a new one erected not long ago by the *St George*—a British scientific expedition. But a few feet back in the brush I found a very old weathered cask with the letters U.S. MAIL still faintly visible. It was the last remaining trace of what was probably the world's most romantic postal service." Robinson's letter followed him shortly afterward, taken on to Tahiti by the *Illyria*, a brigantine on a scientific cruise of the South Seas.

In December 1933, Irving and Electa Johnson and their crew aboard the 92-foot schooner *Yankee* left a letter in "the barrel stuck on top of a pole . . . and many months later found it had worked: a passing yacht had acted as mailman."

John Caldwell, a newly discharged American GI, stopped here in his 29-foot boat *Pagan* in July 1946. Caldwell had left his Australian war bride Mary in Sydney almost a year earlier and, with the immediate postwar scarcity of transportation, had been unable to get back to her since by plane or ship. In desperation, stranded in Panama, he bought *Pagan*, a rundown cutter, and headed across the Pacific without knowing how to sail or navigate. He did manage to reach Post Office Bay and left the bulky letter he'd been writing to Mary, wrapped with a five-dollar bill and a rubber band, in "the parched ornamental barrel" he found on the beach. His voyage had already been hellish, and Caldwell was very uncertain that he'd live to make it across the Pacific to his wife.

> Five dollars from my small remaining funds was a lot; but I wanted Mary to get that letter, and if five dollars would insure it—and I felt it would—then the money didn't matter. I stood by the traditional landmark for a moment wondering if the letter would ever reach Mary. . . .
>
> As I rowed out to *Pagan* I was oblivious to the dismal countenance of the surroundings or the growing cold. My mind was across the Pacific. It was also with the letter in the barrel. An unholy melancholy was on me. I was swept with the futile remorse of great desire, hindered by need of lengthy patience, and burdened by uncertainty.

Caldwell "posted" his letter on July 22 and sailed away the next day. Some vessel picked it up because ten weeks later, in early October, Mary received it in Sydney. It informed her that he hoped to arrive in *Pagan* around the end of September. In other words, he was then at least a week late. But *Pagan* didn't make it. Caldwell survived a hurricane, but shipwrecked on the reefs of the Fiji Islands. He lived, though, to crawl ashore across the rocks and eventually make his way by copra ship, motor bus, and army bomber to Sydney, where he was reunited with Mary on December 3, 1946. (John Caldwell wrote the full story of this adventure in his book *Desperate Voyage*.)

* * *

Watered, wooded, and carrying "thirty large terrapin on board," the *Beagle* dropped the Galapagos Islands astern on October 20 and sailed southwest across the Pacific.

FitzRoy's instructions, beyond the South American survey, were to take only what time he needed to make celestial observations that would establish a chain of consecutive longitude distances around the globe, back to the vessel's starting point at Plymouth—back to that rock in Plymouth breakwater which had been the starting point for all his longitude observations. The rest of the world, after South America, was covered with the thoroughness of a five-day, ten-country bus trip across Europe. The ship spent only ten days in both Tahiti and New Zealand. The speed of the return voyage, the light duty now that surveying was largely behind them, gave the men aboard the relative leisure to view the world as regular tourists.

FitzRoy was astonished at the size of Sydney. "I saw a well-built city covering the country near the port." But he didn't think it would last, and, like many since, he had a low opinion of Australian culture.

It is difficult to believe that Sydney will continue to flourish in proportion to its rise. It has sprung into existence too suddenly. Convicts have forced its growth, even as a hot-bed forces plants, and premature decay may be expected from such early maturity. . . .

There must be great difficulty in bringing up a family well in that country, in consequence of the demoralizing influence of convict servants, to which almost all children must be more or less exposed. Besides, literature is at a low ebb; most people are anxious about active farming, or commercial pursuits, which leave little leisure for reflection, or for reading more than those fritterers of the mind, daily newspapers and ephemeral trash.

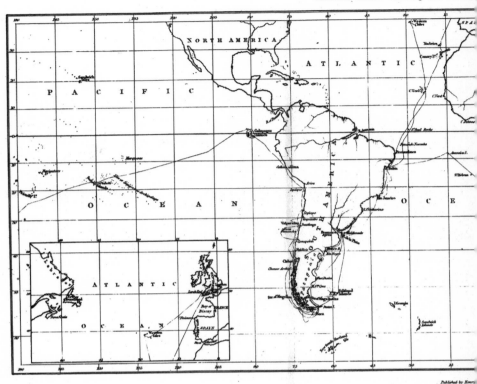

Map from FitzRoy's *Narrative* showing the *Beagle*'s track around the world.
(Narrative of HMS *Adventure* and *Beagle, by Robert FitzRoy*)

The *Beagle* remained in Sydney only two weeks. As the ship moved across the globe toward England, its crew grew hungrier for home. None more so than her captain, on whom the weight of the voyage had taken a visible toll. Phillip Parker King, FitzRoy's commander aboard HMS *Adventure* on the *Beagle*'s first voyage, had moved to Sydney, and FitzRoy called upon him there. King was appalled at the change in his former youthful commander, now still only thirty years old.

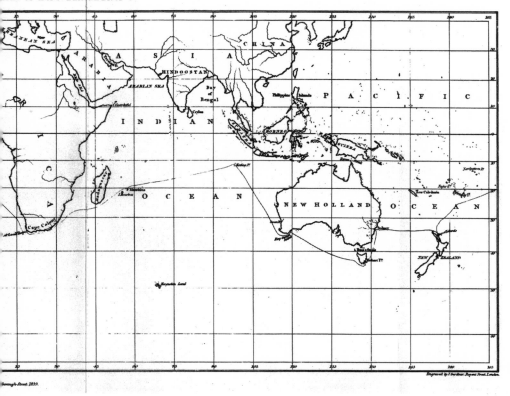

Sydney, 2 February 1836
My dear Beaufort,
    You will have heard from FitzRoy who has been here a
fortnight and sailed on the 30th for Van Diemens Land on his
return. I regret to say he has suffered very much and is yet
suffering much from ill health—he has had a very severe shake
to his constitution which a little *rest* in England will I hope
restore for he is an excellent fellow and will I am satisfied yet
be a shining ornament to our service . . . .
    Very truly yours
    Phillip P. King

And Darwin, in a letter home at the same time, echoed King's concern, and added one more thought.

> From Sydney we go to Hobart Town, from thence to King George Sound and then adios to Australia. From Hobart Town being super-added to the list of places I think we shall not reach England before September: But thank God the captain is as home sick as I am, and I trust he will rather grow worse than better. . . . I have been for the last twelve months on very cordial terms with him. He is an extraordinary but noble character, unfortunately, however, affected with strong peculiarities of temper. Of this, no man is more aware than himself, as he shows by his attempts to conquer them.
>
> I often doubt what will be his end; under many circumstances I am sure it would be a brilliant one, under others I fear a very unhappy one.

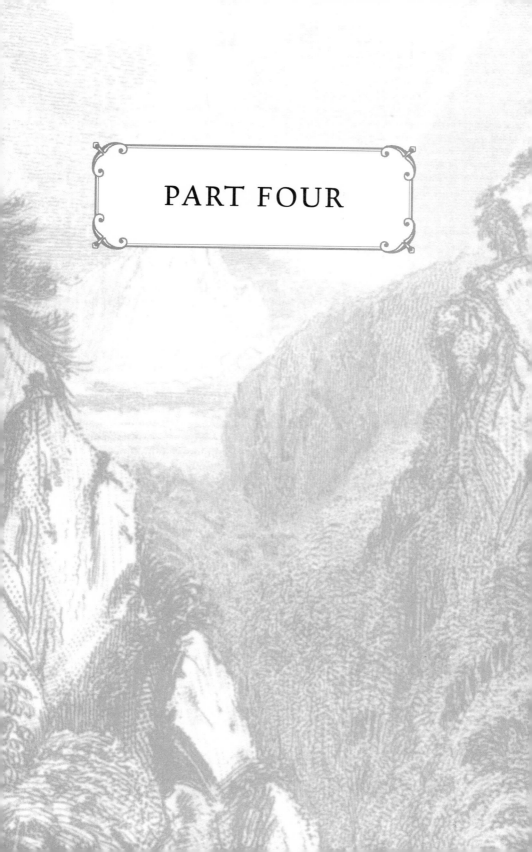

PART FOUR

20

The *Beagle* dropped anchor in Falmouth, England, on October 2, 1836. On her voyage of 58 months, 43 had been spent in South America. From Peru, she had sailed home in just 13 months, all aboard her ready to be done with the voyage, their thoughts fastened upon England and home.

Darwin, who had been seasick through one last gale in the Bay of Biscay as the *Beagle* worked her way up to the Western Approaches, was packed and thoroughly ready. He disembarked at once. That night ("a dreadfully stormy one") he started north from Falmouth by mail coach to Shrewsbury and reached The Mount, the Darwin family home, at breakfast time on Wednesday, October 5.

"Why, the shape of his head is quite altered," said his father to Darwin's gaping sisters. Though neither Darwin nor his father were adherents of phrenology, both seemed to have believed, as FitzRoy did about his Fuegians, that concentration and employment of mental powers could affect the shape of the cranium. Darwin cites this observation by his father, "the most acute observer whom I ever saw," as evidence that "my mind became developed through my pursuits during the voyage." In fact, Darwin was balding prematurely. While away at sea he had lost most of the hair on top of his head.

The *Beagle* was an instant dockside attraction. In Falmouth, newspapers announced her arrival from around the world. People crowded the quays to see her, board her, put their hands upon her as they would today the space shuttle, and to question her crew about storms and savages. Her captain's company was eagerly sought. On October 3, the day after her arrival, FitzRoy was invited to the home of Robert Were Fox, Falmouth's eminent Quaker scientist. His daughter, Caroline Fox, recorded the visit in her journal.

> October 3. — Captain FitzRoy came to tea. He returned yesterday from a five years' voyage, in HMS *Beagle*, of scientific research round the world, and is going to write a book. He came to see papa's dipping needle deflector, with which he was highly delighted. . . . He stayed till after eleven, and is a most agreeable, gentlemanlike young man. He has had a delightful voyage, and made many discoveries, as there were several scientific men on board.

The *Beagle* sailed on to Plymouth and Portsmouth to receive visits from Admiralty bigwigs, "repectable-looking people," (who came aboard by the accommodation ladder) and "others" (humbler sightseers who were permitted to climb into the ship on a rough plank).

Darwin wrote to FitzRoy from Shrewsbury, offering sympathy at finding himself once more in "that horrid Plymouth," where he had languished for so many months before beginning the voyage. But the time had been well spent for FitzRoy, as he revealed in his reply to Darwin.

> Dearest Philos . . . that horrid place contains a treasure to me which even you were ignorant of!! Now guess and think and guess again. Believe it, or not,—the news is true—I am going to be married!!!!!! to Mary O'Brien. Now you may know that I had decided on this step, long, very long ago. All is settled and we shall be married in December.

On top of all the anxieties FitzRoy had suffered through his mission, there had been the question many seamen take with them when they leave a loved one at home: Will she still be there?

The *Beagle* sailed on up the Channel and into the Thames to Greenwich, where she let go her anchor on the zero meridian and FitzRoy made his final observations of the voyage. The ship remained at Greenwich for two and a half weeks for visits by the Astronomer Royal and other guests, then dropped downstream on the tide to Woolwich Dockyard, where she had been built and launched sixteen years before. On November 17 the ship decommissioned, her crew paid off.

Many of the seamen and officers who left the ship and parted from one another had been aboard the *Beagle* on both her voyages, under FitzRoy's command for more than six years. They had faced something very like war during those years; together they had fought the sea, the Fuegians, and the most ferocious weather on Earth, and many times they had saved one another's lives. A few of their number had died. FitzRoy did not write about it, but disbanding beside the *Beagle* on the dock in Woolwich that day would have been as emotional and wrenching for those seamen as the breakup of a tight battalion of long-serving soldiers at the end of a world war. The men went home, or they found berths aboard other ships. Some of the officers went on to notable careers. Most of the crew simply disappear from record.

FitzRoy went home to Onslow Square in London to prepare for his wedding and begin the work of overseeing the drawing and production of new, wonderfully accurate nautical charts from his years of prodigious surveying.

Darwin felt awkward at home. He had left as a boy just gradu-ated from university and come back a grown man, an adventurer, a working scientist. He'd spent years galloping around South America with gauchos and soldiers, eating wild animals over camp fires, trading with natives, roaming through jungles, sharing desperate adventures with tough seamen.

The Mount was filled with sisters. They fussed over him. They expected him now to stay at home and settle into life as a country gentleman, to pick up again the threads of his preparation to be a country parson. But Darwin had been around the world, and while travel no longer held any attraction, the disciplines of the natural sciences, in which he had steeped himself for five years, and the community of his fellow scientists, now beckoned and urged him on to new exploration. He could not go home again.

> Towards the close of our voyage I received a letter whilst at Ascension [Island], in which my sisters told me that Sedgwick had called on my father and said that I should take a place among the leading scientific men. I could not at the time understand how he could have learnt anything of my proceedings, but I heard (I believe afterwards) that Henslow had read some of the letters which I wrote to him before the Philosophical Soc. of Cambridge and had printed them for private distribution. My collection of fossil bones, which had been sent to Henslow, also excited considerable attention among palæontologists. After reading this letter I clambered over the mountains of Ascension with a bounding step and made the volcanic rocks resound under my geological hammer!

Darwin came home to the fruit of his tremendous industry over the course of the last five years. His collection amounted to a full museum's worth of the world's natural marvels: whole groups of plants and insects, birds, small and large animals and reptiles, corals, shellfish and other sea creatures and invertebrates, bones, fossils, rocks, and minerals. His labeling, cataloging, wrapping, drying, and bottling of specimens had been meticulous and thorough.

Everything had gone to Darwin's mentor, Professor Henslow in Cambridge, who unpacked each box and crate on arrival, checked its condition and need for further preservation, and stored it for Darwin's return. Henslow had been unable to resist putting out word of the magnificent collection accumulating, or publishing extracts of Darwin's letters to him. He had spent five years paving the way for Darwin's reappearance, so that in the world of the professors and the burgeoning natural sciences, Darwin was already famous. Charles Lyell wanted to meet him, the Geological Society wanted to elect him a Fellow. Museums everywhere wanted his bones and butterflies. There was no going back, no more search for a career. He had become somebody. All that running away from his studies—beetle collecting, riding, hunting, and shooting—had found him a destiny.

After ten days at home, he escaped to London, to stay with his older brother Erasmus. His life there became frantic with activity—tea parties with the Lyells, irresistible invitations from leading scientists. In December he moved to Cambridge, to the site of his collection. There, with the help of his *Beagle* assistant, Syms Covington, he began to look through everything, to unravel his voyage, to see what he had really done.

He sent his specimens off to the experts who could identify them, determine whether he had found new species, give them their official Latin names. Most of the early excitement was naturally generated by the big fossilized bones of the extinct creature Darwin had dug up on the east coast of Patagonia, the *Megatherium*—or was it a *Scelidotherium*, or a *Toxodon*, or all three? Here was something tangible to wonder at, obviously destined for the museums already clamoring for them. The smaller stuff—the Galapagos birds with their varying sizes of beaks, the different shells of the tortoises—were of subtler interest, and it would be some time before Darwin began to cogitate on just why they varied.

He had a book to write. Late in the voyage, while FitzRoy was beginning to prepare and collate his own journals for publication, he asked Darwin if he could read some of what he had been writing in his journals. FitzRoy thought "Philos's" observations good enough to incorporate into the long narrative of the *Beagle*'s two voyages he was planning to publish. Initially, both Darwin and FitzRoy saw this contribution as sections slipped into the larger work, but later the size and readability of Darwin's diary led them both to believe it could form a distinct volume on the natural history of the countries the ship had visited. Out at sea, at the time of FitzRoy's suggestion, Darwin was flattered, excited by the idea of another book to write (in addition to the book on the geology of the counties visited, which he had first conceived early in the voyage at the Cape Verde Islands). But by late 1836, surrounded by overflowing crates of specimens in Cambridge, his head filled with a kaleidescope of images and ideas, such a book had become a mounting imperative to him. Threads and shapes and anomalies of creation were coursing through Darwin's brain, hot-wiring his synapses, and he felt an urgent need to make sense of it all.

Such a book would also, he knew from the attention he was getting, put him on the map in the scientific community. This was what he wanted now, to take his place, as Sedgwick had suggested, among the leading scientists of the day, men like Lyell, whom Darwin respected and admired hugely, who influenced the thinking of the civilized world. He wanted to be one of them. A book about his voyage would do it.

Darwin began writing it in Cambridge in January 1837. In March, with most of his collection dispersed to the experts for identification, he moved back to London, renting rooms in Great Marlborough Street for himself and Covington who was still working as his general assistant. There he continued work on his book. It closely followed the daily journal he had kept throughout the voyage, on land and sea.

• • •

FitzRoy was at least as busy. After his marriage, he and his wife settled into domestic life in London, and Mary was soon pregnant.

FitzRoy returned to accolades rare for a naval officer in peacetime. He was publically thanked in Parliament. The Royal Geographic Society presented him with their gold medal. The appreciation of the Admiralty—for him more important than any honor—was deep. The quality of his work as a marine surveyor was immediately evident. It was so thorough and accurate that the resulting charts were used for more than a century.

In aristocratic, social, and scientific circles, FitzRoy was famous—much more so than Darwin, whose renown was narrowly confined to the community of academics and naturalists. On his return to London FitzRoy was much sought after: a dashing, witty, charming officer, a gentleman, seaman, and scientist of extraordinary accomplishments, with equally extraordinary tales to tell. More than anything, he had sailed around the world, and the neat geometry of this feat provided the shape and allure to all he had done. He was England's nineteenth-century astronaut returned to Earth. Everyone wanted to meet and talk with Captain Robert FitzRoy.

When his work on his surveys and charts was completed, FitzRoy didn't seek another sailing commission. Despite the inroads he had made in his personal fortune he was still financially independent, and with Mary pregnant and his own book of the voyage to write, he had more than enough to keep him at home.

The narrative he had long planned to publish consisted of two main parts. A first volume would cover the voyages of the *Adventure* and the *Beagle* during the years 1826–1830 (with Pringle Stokes and FitzRoy as successive captains of the *Beagle*), when both ships had been under the overall command of Captain Phillip Parker King; a second volume for the *Beagle*'s five-year voyage, from 1831 to 1836. Volume One was to be authored by Captain King, but as he had moved to Australia, the production of this fell entirely to FitzRoy, who had to put together and edit the first book from his own, Stokes's, and King's thick pile of notes and logs.

Volume Two, FitzRoy's own dense day-by-day account of the second voyage, was enormous, if not quite epic. With lengthy essays on the state of native peoples, descriptions of coastlines, weather, sea conditions, shiphandling, adventures ashore, and, not least, the repatriation of the Fuegians, it would run to 695 pages and a quarter of a million words when completed. Multiple appendixes covered 350 pages in a separate volume. Darwin's book would form a third volume, but apart from that it was all FitzRoy's to do. He began early in 1837.

It must have seemed to him at first an enjoyable task: to stay at home with his wife and coming child, to voyage daily no farther than from bedroom to study. To suffer no setbacks from storms but rather to sit in a peaceful room in England and view the weather from the dry side of a windowpane—to a seaman such a secure berth holds an intense appreciation. To be free of the bedevilment of hostile natives. To light a fire, trim a wick, and spread around him at his desk the notebooks, logbooks, maps, and drawings from which he could select and produce a coherent account; to draw up a good chair, dip a smooth nib into a still inkwell, and begin.

He was swamped. Deciphering the handwritten logs of two other men, King and Stokes, going over their every word, checking each estimation of time or distance for mistakes or inconsistencies—it would be their accounts but his name would testify to it all—proved the most serious drudgery. This was not writing, not the creative effort that returns some glow of satisfaction for long hours in a chair.

He dined out, spent time with Mary and his new daughter, took walks—from his house in Onslow Square, which faced a garden, it was a ten-minute stroll to the Royal Geographic Society, or Hyde Park, a short cab ride from Admiralty House, Whitehall, Mayfair, or his club in St. James—but always the enormity of the work to be done gnawed in his mind and pulled him back to his study. It pulled him out of sleep with a jolt in the predawn hours, and it kept him bent over his desk late into the night. The two volumes, and a third comprised of essays and appendixes, amounted to 1,650 pages and over half a million words (six times the size of this book), and none of it could be done by anybody else. He still had his servant from the *Beagle*, Fuller, whom he'd recruited to be his clerk after Hellyer's death, to assist him, to run errands, to help arrange for the engravings of plates for illustrations, but no assistant or clerk could check what only FitzRoy knew. The work pulled FitzRoy's high tension wires tighter.

He and Darwin saw each other rarely after the voyage. Even when both were living in London, they seemed to grow farther apart.

> I saw FitzRoy only occasionally after our return home [Darwin wrote many years later], for I was always afraid of unintentionally offending him, and did so once, almost beyond mutual reconciliation.

In the spring of 1837, Darwin came to tea with the FitzRoys. "So very beautiful & religious a lady," he thought Mrs. FitzRoy. But the visit was probably less for social reasons than to confer about specimens both had taken from the Galapagos Islands: Darwin needed FitzRoy's help to identify his birds.

It was not a happy reunion. Darwin came away bristling over FitzRoy's ill humor, firing off several letters about his former messmate.

> The Captain is going on very well [he wrote to his sister], — that is for a man, who has the most consummate skill in looking at everything & every body in a perverted manner.

And he wrote to Charles Lyell, with whom he had now become close friends.

> I never cease wondering at his character, so full of good & generous traits but spoiled by such an unlucky temper. — Some part of his brain wants mending: nothing else will account for his manner of viewing things.

Nevertheless, FitzRoy gave him what he asked for.

Darwin had sent his bird, mammal, reptile, and insect specimens to the London Zoological Society for identification. One of its fellows, John Gould, undertook the classification of his birds. Gould was a widely respected ornithologist and taxonomist, so Darwin had to believe what Gould told him about his Galapagos collection: three mockingbirds Darwin had collected from three different islands were not variations of South American birds but three new and distinct species. A number of birds Darwin had labeled variously as finches, wrens, and blackbirds, were all finches, Gould claimed, of a new group unknown beyond the Galapagos Islands. Their very different beaks made them finches of different species, and Gould believed that, like the mockingbirds, each species came from a different island.

Or so it appeared, but here Darwin's collecting had been uncharacteristically clumsy. He had commingled specimens from several islands, never imagining that different but identical islands, not far apart, might produce different birds. Gould needed more Galapagos finches, properly labeled. Darwin could hardly return to the Pacific, so he had approached FitzRoy to borrow birds the captain, his assistant Fuller, and various seamen had clubbed and taken away from the islands, which FitzRoy had already presented to the British Museum. Despite their quarrel over tea, FitzRoy had two sets of bird skins sent to Darwin. Syms Covington had his own four birds from the Galapagos. All these were examined by Darwin and Gould and com-

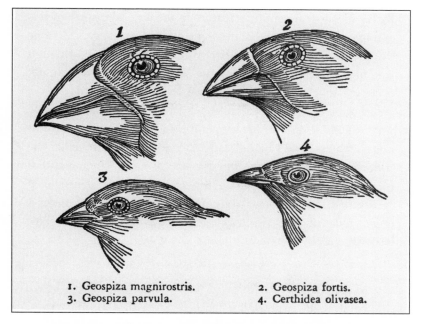

1. Geospiza magnirostris.
2. Geospiza fortis.
3. Geospiza parvula.
4. Certhidea olivasea.

One genus; different species. The finches with varying beaks that Darwin found at the Galapagos Islands. (Narrative of HMS *Adventure* and *Beagle, by Robert FitzRoy*)

pared with Darwin's lists and catalogs, in an effort to confirm what Gould claimed: different islands, different species.

Another fellow from the Royal Zoological Society, Thomas Bell, who had been identifying Darwin's reptiles, came back with a parallel conclusion: each island of the Galapagos chain had produced its own distinct species of iguana lizard.

Darwin received this news in the spring of 1837 while he was deep in the writing of his book about the voyage. It concurred with a remark the islands' vice governor had made to him while he was there eighteen months earlier, something about the markings on the shells of the islands' tortoises. He hadn't taken much notice at the time, but now the memory of it pealed through his mind. He put it all together in his chapter on the Galapagos Islands.

By far the most remarkable feature in the natural history of this archipelago [is] that the different islands to a considerable extent are inhabited by a different set of beings. My attention was first called to this fact by the Vice-Governor, Mr Lawson, declaring that the tortoises differed from the different islands, and that he could with certainty tell from which island any one was brought. I did not for some time pay sufficient attention to this statement, and I had already partially mingled together the collections from two of the islands. I never dreamed that islands, about 50 or 60 miles apart, and most of them in sight of each other, formed of precisely the same rocks, placed under a quite similar climate, rising to a nearly equal height, would have been differently tenanted.

On March 14, while Darwin was preoccupied with identifying his finches, he attended a lecture given by Gould at the Zoological Society, on the subject of the South American *Rheas*, or ostriches, specimens of which Darwin had brought back and presented to the society. In northern Patagonia, ostriches were common. The gauchos who ate them had told Darwin of a similar but much rarer bird they called the *Avestruz petiso* ("little ostrich"). Darwin had looked for this bird with no success, until farther south, at Port Desire in January 1834, he was eating what he thought was an ostrich, shot by one of the *Beagle*'s company, when he noticed it was smaller. The meal was over and the bird fully consumed by the time Darwin realized it was an *Avestruz petiso*. But rooting through the spat-out bones, skin, feathers, and leftovers, he came up with "a very nearly perfect specimen" that was later exhibited at the Zoological Society. But it was not the rarity he had been led to believe.

Among the Patagonian Indians in the Strait of Magellan, we found a half Indian, who had lived some years with the tribe, but had been born in the northern provinces. I asked him if he had ever heard of the Avestruz Petise? He answered by saying, "Why, there are none others in these southern countries."

Gould did Darwin the honor of naming the smaller, southerly ostrich *Rhea darwinii*, after the man who had eaten it. The point of his lecture was that the *Rhea darwinii* was just as common in southern Patagonia as the bigger bird was farther north, yet there were differences enough between the two to classify them as different species. Different places, different birds, though of essentially the same feather. Just like the finches.

Why, Darwin began to wonder, had the creator bothered with such subtle differences? Why not make one species of finch and let it suffice for one small group of islands? Why make a baker's dozen? Why two ostriches where one would do?

A little over a year earlier, on a sunny afternoon in January 1836, while out hunting kangaroo in New South Wales, Australia, such questions hadn't troubled him.

I had been lying on a sunny bank & was reflecting on the strange character of the Animals of this country as compared with the rest of the World. An unbeliever in every thing beyond his own reason, might exclaim "Surely two distinct Creators must have been [at] work; their object however has been the same & certainly the end in each case is complete." — Whilst thus thinking, I observed the conical pitfall of a Lion-ant: — A fly fell in & immediately disappeared; then came a large but unwary Ant; his struggles to escape being very violent, the little jets of sand . . . were promptly directed against him. His fate however was better than that of the poor fly's: — Without a doubt this predæcious Larva belongs to the same genus, but to a different species from the European one. Now what would the Disbeliever say to this? Would any two workmen ever hit on so beautiful, so simple & yet so artificial a contrivance? It cannot be thought so. — The one hand has surely worked throughout the universe. A Geologist perhaps would suggest, that the periods of Creation have been distinct & remote the one from the other; that the Creator rested in his labor.

Darwin did not, in January 1836, question why the creator had bothered to make two different versions of the same insect. Like most nineteenth-century scientists and thinkers, he did not question that a creator had been behind it all. Time, he suggested, like a good Lyellian geologist, had simply made the creator rethink his model. It didn't occur to Darwin then that the place, which after all looked not unlike England, might have something to do with it.

Fourteen months later, however, he *was* questioning the creator's efforts. Nevertheless he transcribed this diary entry almost exactly into the book he was writing—his questions had not yet led him to a solid conclusion. (By 1845, when the second, revised, edition of his book was published, that conclusion had come, and Darwin relegated this incident with the lion ant to a footnote, shorn entirely of his ruminations about what a disbeliever might think of the creator's scheme of things. By then he had become, in his own words, "an unbeliever in every thing beyond his own reason.")

Darwin's reasoning brought him to an inescapable idea: perhaps the creator had not created, at the beginning of time—on day five according to the Book of Genesis—all those different finches and placed them on the Galapagos. It seemed more reasonable that the islands might have been populated by South American birds, which, in time, could have adapted to the peculiarities and food sources on each island, until they became so distinctly, consistently different that they had become a new species. The same was possible with the South American ostriches. One species might have evolved from another.

It wasn't a new idea, and certainly not to Darwin: his own grandfather had been an evolutionist. The Frenchman Lamarck had suggested the same thing almost forty years earlier, but in far more controversial terms: that humans had evolved from apes. Lamarck believed each plant, each animal, contained a "nervous fluid" that enabled it to generate in new directions, adapting to its local environment. The ancestors of giraffes, he suggested, extended their necks over time by stretching to eat the leaves on overhead trees, causing their nervous fluid to flow into their necks which, over successive generations, grew longer. Apes might similarly have dropped to the ground from trees and found that walking or running upright was the more efficient posture.

However clever or reasonable an idea this might have seemed, it was religious heresy. It suggested a godless world propelled by ungoverned, earthly forces. Such an anarchy of creation would produce a nightmare world of incessant haphazard mutation, of gargoyle monsters. And that could not be so, people believed, for the order and beauty of the world was everywhere apparent.

But Darwin saw a middle way: the transforming of one species into another, a process guided by nature, not by God, whereby a plant or animal gradually selected the physical peculiarities best adapted to its environment.

Inescapably with this came the corollary that Adam and Eve were not the semi-angels created by God in the Garden of Eden, but animals descended from other animals, most closely and recently from monkeys. Nothing encouraged this notion more than Darwin's acquaintance with Fuegians. "Viewing such men, one can hardly make oneself believe that they are fellow creatures placed in the same world," he had written in his voyaging diary.

He had found it difficult to believe that he and they were one species. Bent, hulking, primitive, living in shelters far less complex than a bird's nest, the Fuegians had seemed to provide a picture of Man at the dawn of his transformation from ape, perhaps closer to the root they had sprung from than to the apotheosis represented by European man. At the Zoological Society in London he had observed an orangutan named Jenny throwing a tantrum when her keeper would not give her an apple.

She threw herself on her back, kicked & cried, precisely like a naughty child. — She then looked very sulky & after two or three fits of pashion, the keeper said, "Jenny if you will stop bawling & be a good girl, I will give you this apple. —" She certainly understood every word of this, &, though like a child, she had great work to stop whining, she at last succeeded, & then got the apple, with which she jumped into a chair & began eating it, with the most contented countenance imaginable.

Remembering the relentless demands of "yammerschooner," Darwin went home and wrote in his notebook: "Compare the Fuegian & Ourang outang & dare to say difference so great."

Three years earlier, in Tierra del Fuego, he had written in his diary, "Nature, by making habit omnipotent, has fitted the Fuegian to the climate & productions of his country," already unconsciously anticipating his new thinking about the "transformism," as he first thought of it, of species.

He wasn't afraid of what he was thinking. From all he had seen on his five-year circumnavigation, it made sense. The scientist in him embraced it. However, he was well aware of its religious impropriety. It was after all blasphemy, and Darwin was not a bold man in his researches. But he was conscientious. He had no wish to be controversial, so he kept his musings to himself. That summer of 1837 he started scribbling in notebooks, jotting down and organizing such thoughts. It would be twenty-two years before he published his ideas about transmutation, and then he did so only because another naturalist, a malaria-ridden Englishman wandering through Borneo, shocked him into action by sending him a letter suggesting identical conclusions.

Darwin finished his account of the voyage in September 1837, after nine months of concentrated transcription and editing of his own journals. It contained no allusion to his ideas on the transmutation of species. It was simply a fabulously adventurous travel book, filled with the descriptions of a teeming natural world by a remarkably observant scientist. There was nothing in it to offend anybody. He sent it off to the publisher, Henry Colburn, whose office was conveniently close by, on Great Marlborough Street in London where Darwin himself was living. The finished narrative, to be included with the three volumes FitzRoy was preparing, was due to be published the following year, 1838, but Colburn's inquiries to FitzRoy about the date of his completion met with evasion.

FitzRoy was desperately trying to pull together a mass of chronologically overlapping, generally uninspiring seafaring accounts by three different authors into some sort of readable shape. He was brittle and ready to snap.

He finally did so when Darwin sent him the preface to his account in November.

FitzRoy reacted to it with a cold fury. Darwin's preface, he wrote back to him, showed a shameful lack of acknowledgment of all those aboard the *Beagle*—not only the captain himself but the ship's officers—who had tirelessly and generously assisted Darwin, looked after him, gave up room for him, put his well-being before their own on all occasions, and generally made possible all his efforts throughout the voyage, and of course, subsequently, the publication of his book. FitzRoy was right. Darwin had offered little thanks to the men who had nursed him through seasickness and looked after him as a favored guest for five years. There was no distinction made between Beaufort's kindness in allowing Darwin to go on the voyage and FitzRoy's unbounded generosity to him both as a guest and with the facilities of the Admiralty in shipping home his collection.

Darwin apologized. He rewrote his preface in more generous terms.

But there was more than slighting acknowledgment and offended feelings to their growing estrangement. Something else was happening to drive the two men irreconcilably far apart. Within two years of returning to England, both experienced profound changes of thought that were in direct and contentious opposition.

While Darwin was moving ineluctably toward scientific enlightenment, FitzRoy was heading fast in the other direction.

21

Sometime between coming home in October 1836 and finishing his narrative two and a half years later—perhaps under the influence of his new wife—FitzRoy discovered, or rediscovered, an ardent faith in God and in the literal word of the Bible, which he had begun to reread with the closest of attention. This would bring an end to any warmth left in his relationship with his former shipmate.

Perhaps Darwin had mentioned his thoughts about transforming species when they'd taken tea together. This was more than likely after their many years of scientific discussion and debate over shared meals. Such thoughts would have drawn a ferocious response from the captain and sent Darwin away musing angrily about an ill-humored crank.

FitzRoy read Darwin's book. Its inferential trend—all those proliferating species—even without conclusions, would have been quite clear to him. He found it abominable, full of heresy.

The only quadruped native to the island is a large wolf-like fox which is common to both East and West Falkland. I have no doubt it is a peculiar species, and confined to this archipelago . . .

This clearly implied the sort of natural, rather than God-designed, transmutation that had infuriated FitzRoy over tea. Having invited him to write a book, and made arrangements with their publisher, FitzRoy could hardly now stop Darwin or edit him. But he could oppose Darwin's views. So he went back into his own book to argue with him.

The only difference between the Patagonian and the Falkland foxes, FitzRoy declared, was their size and thickness of coat, and this he could readily explain.

> Naturalists say that these foxes are peculiar to this archipelago, and they find difficulty in acounting for their presence in that quarter only. That they are now peculiar cannot be doubted; but how long they have been so is a very different question. As I know that three hairy sheep, brought to England from Sierra Leone in Africa, became wooly in a few years, and that wooly sheep soon become hairy in a hot country (besides that their outward form alters considerably after a few generations); and as I have seen and heard of wild cats, known to have been born in a domestic state, whose size surpassed that of their parents so much as to be remarkable; whose coats had become long and rough; and whose physiognomies were quite different from those of their race who were still domestic; I can see nothing extraordinary in foxes carried from Tierra del Fuego to Falkland Island becoming longer-legged, more bulky, and differently coated.

Darwin of course was suggesting nothing more than this sort of natural adaptation. But he termed the later models new species, and here FitzRoy, and countless others, took virulent issue with him and with the implications of such evolutionary thinking. Foxes and birds might change a bit, but men did not grow from monkeys.

After two and a half years sorting through and preparing his two volumes and many appendixes, delaying publication of the multivolume book by a year, then finally finishing, FitzRoy now added two more chapters to his Volume Two. They were his rebuttal to Darwin's unwholesome line of thought.

Chapter 27, titled "Remarks on the Early Migrations of the Human Race," began:

> Having ended my narrative of the Beagle's voyage, I might lay down the pen: but there are some reflections, arising out of circumstances witnessed by myself, and enquiries since made respecting them, that I feel anxious to lay before those who take an interest in such subjects. . . .

He went on to discuss the "various countenances, heads, shapes, sizes, colours, and other peculiarities of the human race." He provided a table, "Castes arising from the mixture of European, Indian, and Negro," of at least twenty-three "distinct varieties of the human race" which he observed in the city of Lima, Peru: White, Creole, Indian, Mestiso, Negro, Mulatto, Quateron, Zambo . . .

> Having seen how all the varieties of colour may be produced from white, red, and black, we pause, because at fault, and so we should remain, did we rely on our own unassisted reason. But, turning to the Bible, we find in the history of those by whom the earth was peopled, after the flood, a curse pronounced on Ham and his descendents; and it is curious that the name Ham should mean "heat-brown-scorched," while that of Cush his son, means "Black;" that Japheth should imply "handsome," and that Shem, from whose line our Saviour was descended, should mean "name — renown — he who is put or placed."

The Bible made it clear, FitzRoy wrote, that the cursed Negro descendents of Cush ended up as the black, red, and brown "aborigines" of the southerly world of Africa, Australia, and the Pacific islands, while the handsome, renowned, and white descendants of Shem and Japheth made their way to northern Europe.

FitzRoy's genealogy presses on with unquestioned sure-footedness and an avid racial speculation that wouldn't be out of place in pamphlets published by today's white Aryan brotherhoods.

> It is likely that some of Abraham's bond-women were either black or mulatto, being descendents of Ham; perhaps of Cush: and it is hardly possible that Hagar should not have been dark, even black, considering her parentage; in which case Ishmael would have been copper-coloured, or mulatto, and some, if not all of Abraham's sons by concubines, would have been of those colours.

These descendants of Noah, once civilized, began a migratory spread across the globe—FitzRoy limns this movement adeptly as a seaman, making the same case Thor Heyerdhal did with *Kon Tiki* for the drift of primitive man across the oceans on log and reed craft. He finds a deeper, surer root of ancestry for his voyagers than Heyerdhal: "No one . . . can read about those countries . . . without being struck by the traces of Hebrew ceremonies and rites . . . scattered through the more populous, if not through all the nations upon earth."

But, except for that fortunate branch of the Shem-Japheth clan that forked west and north and finally turned into Englishmen, FitzRoy reasoned that these wanderers became increasingly primitive, falling farther from grace with the distance and length of time they roamed from the influence of holy precincts.

> Let us suppose, for illustration, that a party of men and women left Asia Minor in a civilized state. Before they had wandered far, no writing materials or clothes would have remained . . . and their children would have been taught only to provide for daily wants, food, and perhaps some substitute for clothes, such as skins. Their grand-children would have been in a still worse condition as to information or traces of civilization, and in each

succeeding generation would have fallen lower on the scale, until they became savages in the fullest sense of the words; from which degraded condition they would not rise a step by their own exertions.

None in all the wandering, forgetful tribes of Israel had fallen lower than the Fuegian, the human nadir at the bottom of the trajectory, whom FitzRoy had tried so hard to raise up again.

So FitzRoy approached his main argument—his rebuttal to what would, in time, be called Darwinism.

That man could have been first created in an infant, or a savage state, appears to my apprehension impossible; . . . because — if an infant — who nursed, who fed, who protected him till able to subsist alone? and, if a savage, he would have been utterly helpless. Destitute of the instinct possessed by brutes, — with organs inexperienced (however perfect), and with a mind absolutely vacant; neither his eye, his ear, his hand, nor his foot would have been available, and after a few hours of apathetic existence he must have perished. The only idea I can reconcile to reason is that man was created perfect in body, perfect in mind, and knowing by inspiration enough for the part he had to perform; — such a being it would be worse than folly to call savage.

Have we a shadow of ground for thinking that wild animals or plants have improved since their creation? Can any reasonable man believe that the first of a race, species, or kind, was the most inferior?

In his final chapter, FitzRoy summoned all his circumnavigating experience and common sense to refute the Lyellian view of geology and the doctrines of transmutation that went with it, which directly contradicted the Book of Genesis. Chapter 28 is titled "A Very Few Remarks with Reference to the Deluge."

I suffered much anxiety in former years from a disposition to doubt, if not disbelieve, the inspired History written by Moses. I knew so little of that record, or of the intimate manner in which the Old Testament is connected with the New, that I fancied some events there related might be mythological or fabulous, while I sincerely believed the truth of others; a wavering between opinions, which could only be productive of an unsettled, and therefore unhappy, state of mind. . . .

Much of my own uneasiness was caused by reading works written by men of Voltaire's school; and by those of geologists who contradict, by implication, if not in plain terms, the authenticity of the Scriptures; before I had any acquaintance with the volume they so incautiously impugn. . . . For men who, like myself formerly, are willingly ignorant of the Bible, and doubt its divine inspiration, I can only have one feeling—sincere sorrow. . . .

While led away by sceptical ideas, and knowing extremely little of the Bible, one of my remarks to a friend, on crossing vast plains composed of rolled stones bedded in diluvial detritus some hundred feet in depth, was "this could never have been effected by a forty days' flood."

The "friend" FitzRoy had confided his doubting remarks to had been Darwin. It was April 1834; they were exploring in the whaleboats, sailing inland from the Atlantic coast up the Santa Cruz River in southern Argentina. The river cut across an austere, flat, brownish-yellow Patagonian plain that was evidently comprised of "sea-worn" smooth stones, which FitzRoy and Darwin found embedded 100 feet below the cliff tops of the river's high southern banks. This seemed to prove the truth of Charles Lyell's claims that the geology of the earth had evolved gradually, over an unimaginably long period of time. Here were signs of an eon of inundation and sea action wearing small and smooth the rocks of a sea bottom, now long exposed since the withdrawal of water. It did not seem possible, as FitzRoy had said, that such an effect could have been produced by a temporary flooding.

Yet four years later he was convinced of the callowness of his earlier opinion. It was the ubiquitous seashell, beloved of geologists to prove their point, that FitzRoy used to prove his. If land had sunk slowly over thousands of years, he wrote, to be covered by the sea and then, just as slowly rise again above it, seashells would only be found beneath layers of sediment, mud, and rock strata. But FitzRoy and Darwin had found the cliffs along the Santa Cruz River to be entirely composed from top to bottom of a well-blended mix of earth and shells. This, he reasoned, was the result of water "rushing to and fro, tearing up and heaping together shells which once grew regularly or in beds . . . and from those shells alone my own mind is convinced, (independent of the Scripture) that this earth has undergone an universal deluge." Here was proof geological.

The flood came handily to his aid in dealing with the problem of the extinction of the dinosaurs. His explanation of the modern absence of the largest animals—dinosaurs—whose bones were then being unearthed around the world (including the bones of giant rhinoceroses, elephants, and saber-tooth tigers that had been discovered a few years earlier by workmen laying the foundations of what was to become Trafalgar Square), is perfectly understandable.

Proud man would, in all probability, have despised the huge construction of Noah, and laughed to scorn the idea that the mountains could be covered, even when he saw the waters rising. Thither, in his moral blindness, he would have fled, with numbers of animals that were excluded from the ark, or did not go to it; for we do not see all animals, even of one kind, equally instinctive. As the creatures approached the ark, might it not have been easy to admit some, perhaps the young and small, while the old and large were excluded?

What sound reasoning. Commonsense housekeeping would dictate that Noah and his boys stop the huge, the too troublesome, the undomesticable, and the most ferocious animals at the gangplank.

Dealing with some of the more commonly voiced objections to a flood, FitzRoy used the observations of the astronomer and mathematician Sir John Herschel (who later vehemently disputed the possibility of evolution and stated that the creation of species was an act of God, the "mystery of mysteries"):

Questions arise, such as 'Where did the water come from to make the flood; and where did it go to after the many months it is said to have covered the earth?' . . . Connected with these questions respecting the additional quantity of water is the reflection that the amount must have been very great. This may be placed in another light. Sir John Herschel says, 'On a globe of sixteen inches in diameter . . . a mountain (five miles high) would be represented by a protruberance of no more than one hundredth part of an inch, which is about the thickness of ordinary drawing paper.' Now as there is no entire continent, or even any very extensive tract of land, known, whose general elevation above the sea is anything like half this quantity, it follows, that if we would construct a correct model of our earth, with its seas, continents, and mountains, on a globe sixteen inches in diameter, the whole of the land, with the exception of a few prominent points and ridges, must be comprised on it within the thickness of thin writing paper; and the highest hills would be represented by the smallest visible grains of sand. Such being the case, a coat of varnish would represent the diluvial addition of water; and how small an addition to the mass does it appear!

How wide was the gulf between Darwin and FitzRoy. Darwin stood at the threshold of an expansion of thought and science that would not be equaled for a hundred years, until the technological impetus of World War II incubated the atomic age. FitzRoy in his way was no less a scientist; he read widely and with great discipline, and he applied what he had read with an elastic reason to an epic voyage of discovery. But he was stuck, deeply, by prejudice and the cleaving to an old order, to a mindset a thousand and more years old, when science was subservient to religion. That order was about to be toppled, and the constructs of the Bible smashed like an old wooden bridge, weakened by rot, before the torrent of a spring flood.

It was the last moment in history when the thicknesses of drawing and writing paper could be comfortably used to explain the mysteries of life on Earth.

22

The book—in three volumes: King's, FitzRoy's, and Darwin's, with a fourth of appendixes; hardbound in dark blue-gray cloth—was published by Henry Colburn in the early summer of 1839. The full, ponderous title ran:

NARRATIVE
OF THE
SURVEYING VOYAGES
OF HIS MAJESTY'S SHIPS
ADVENTURE AND *BEAGLE*
BETWEEN
THE YEARS 1826 AND 1836
DESCRIBING THEIR
EXAMINATION OF THE SOUTHERN SHORES
OF
SOUTH AMERICA
AND
THE *BEAGLE'S* CIRCUMNAVIGATION OF THE GLOBE
IN THREE VOLUMES

Darwin's Volume 3 was inauspiciously titled *Journal and Remarks, 1832–1836*. The volumes could be purchased separately, and Darwin's fast became a best-seller, requiring Colburn to reprint it in August, this time giving it a new and grander title: *JOURNAL OF RESEARCHES into the GEOLOGY and NATURAL HISTORY of the Various Countries Visited by H.M.S. Beagle Under the Command of Captain FitzRoy, R.N. From 1832 to 1836.* It was subsequently reprinted many times. Darwin revised it for an 1845 edition brought out by John Murray, Lyell's publisher, and this became, for at least a century, the most widely published and reprinted version. In time it acquired a third, more lasting title, *The Voyage of the Beagle.* It remains a steady seller, one of the world's great travel books.

The rest was dull stuff, except to navigators or armchair sailors keen for the minutiae of the ships' wearyingly repetitive movements on an exhaustive surveying mission. For modern sailors, King's, Stokes's, and FitzRoy's descriptions of wind, wave, and anchorages hold great value as early navigational records, and as curiosities. And FitzRoy's account of his dealings with the Fuegians has always made fascinating reading. But in 1839, as today, it was Darwin's book that was bought and talked about. The other volumes were not then and have rarely since been reprinted. Complete sets of all four volumes were purchased by a few gentlemen, naval officers, travel enthusiasts, clubs and libraries, and after a quick thumb-through, the other three volumes were shelved away.

Reviewers quickly singled out Darwin for praise: a "first-rate landscape-painter with a pen"; "a strong intellectual man and an acute and deep observer." FitzRoy's long-labored-over, dense account of seafaring was generally passed by with the briefest mention. Only his remarks on geology and biblical history drew much response, and this was unfavorable and withering: "On this subject the gallant captain has got quite beyond his depth. . . ."

Ironically, most of the general public, the God-fearing, churchgoing readers who held an unquestioning belief in the biblical account, would have agreed with FitzRoy. They would have been comforted by his interpretation of geology and anthropology, embraced such a scientific corroboration of the Holy Word. But the general public didn't buy, read, or review such books. They read popular and sensational novels, mostly by installment in magazines, religious tracts, newspapers, pamphlets, and the Penny Cyclopaedia. The qualified opinions settling and hardening on both Darwin's and FitzRoy's efforts (Charles Lyell's response to FitzRoy's last chapter, "it beats all the other nonsense I ever read on the subject," was typical) were those of a tiny, smugly enlightened minority, a nearly vaporous film floating on the upper surface of society. But they were the only opinions that counted. They determined the rank and importance of the authors.

This was the moment when all FitzRoy's talents, energies, ambitions, and accomplishments began to be swallowed up by what would increasingly become his singular place in history: as the man who took Charles Darwin around the world, who provided Darwin with the vehicle for his ideas—and in whose shadow he would forever remain a footnote.

He could not have had any idea of the depth and extent of that shadow, but he felt eclipsed, and unappreciated. Darwin's book got all the attention; it sold more, was reprinted, its author was talked about. FitzRoy's book was ignored or, when noted, ridiculed.

He did not have the temperament or the mental equilibrium to navigate through such a devastation.

There is no record of FitzRoy's activity for the next twenty-four months. He did not write, he did not seek another sea commission, he completed no work. Some men, recently married, new to fatherhood, of independent means, might have enjoyed a period of rest after years of voyaging, followed by several years more of intense writing. FitzRoy did not. He had no inclination to idleness—he'd been driving himself relentlessly toward a very personal interpretation of excellence since entering the Royal Naval College at the age of twelve, twenty-two years earlier. Achievement, duty, a deep roving curiosity, and an equally deep but blinkered intelligence, characterized his nature. Relaxation was alien, even unhinging, to him. The two-year gap is conspicuous.

In Chile, when reprimanded by the Admiralty for overstepping his orders, he had suffered a breakdown, "a morbid depression of spirits, & a loss of all decision and resolution," severe enough to remove himself from command. He had been afraid then "that his mind was becoming deranged (being aware of his hereditary disposition)."

That disposition, from his suicidal uncle on his mother's side, was very real, giving him a bottomless inheritance of mercurial mood swings, depression, paranoia, and suspicion of slights, that left him no cushion to absorb life's setbacks. With his big book finished, his conclusions tittered at, glory bestowed on his younger and vastly less accomplished friend, FitzRoy spun into two years of drowning blackness.

He was rescued, again, by an uncle. In the summer of 1841, the Whig Prime Minister Lord Melbourne resigned, and a general election followed. Sir Robert Peel campaigned to bring the Tories back into power, and Lord Londonderry, Peel's friend and FitzRoy's uncle (elder brother of Viscount Castlereagh, FitzRoy's uncle who had slashed his own throat), offered his nephew the candidature of a parliamentary seat for County Durham in place of the retiring Tory MP. Whether FitzRoy had sought or even considered such a position, he grabbed at this life buoy and decided to run.

Durham was a two-seat constituency, Whig and Tory—or Liberal and Conservative as the parties were beginning to call themselves—and FitzRoy appeared to be a shoe-in. Londonderry was a local landowner and feudal loyalties still prevailed: Londonderry's man would normally get his tenants' votes.

But a second Conservative candidate decided to run. Twenty-six-year-old William Sheppard wasn't much of a threat to a Londonderry-backed nominee, and at first he and FitzRoy assumed a friendly, gentlemanly rivalry. Sheppard later wrote, in the effusive idiom of the day, that they spent time together "in unreserved intimacy."

Then a Londonderry tenant told somebody he'd been directed by his landlord to vote for FitzRoy, and the situation turned ungentlemanly. FitzRoy heatedly denied any knowledge of such an attempt at persuasion, but Sheppard didn't believe him. Rather than fight FitzRoy over their seat about this, Sheppard withdrew his candidacy, but he let it be known why. FitzRoy was infuriated, and in a speech in Durham, he called Sheppard's behavior "disgraceful . . . an event unparalleled in the annals of electioneering."

Acrimonious letters flew back and forth between the two antagonists. Both soon claimed that their honor had been impugned. They nominated seconds and demanded satisfaction. Dueling days were almost over, but not quite among the aristocracy: Lord Cardigan had dueled with and wounded an opponent within the last year, and been acquitted of any crime. The Duke of Wellington had fought a duel twelve years earlier. And a demand for satisfaction from the hot-headed FitzRoy had to be taken seriously: his melancholic uncle, Viscount Castlereagh, had fought two duels before ultimately dispatching himself. The notion that political bickering would evolve into a meeting at dawn with pistols was unlikely, but the language exchanged, and form, demanded procedure along certain lines.

The seconds met and discussed terms, which mainly revolved around the wording of conciliatory letters of clarification of what had been said and meant that would be agreeable to both principles. Accord was almost reached, but then fell apart. FitzRoy was too touchy, too outraged, and in Sheppard he met his match for sensitivity and sense of persecution. Sheppard denounced him as "a liar and a slanderer . . . a coward and a knave." Nothing was resolved. FitzRoy was elected. Londonderry advised him to let the matter go, and he did. There it rested until August 25, when FitzRoy came out of a gentleman's club in the Mall and found Sheppard waiting for him with a whip.

"Captain FitzRoy!" shouted the aggrieved loser. "I will not strike you. But consider yourself horsewhipped!"

FitzRoy attacked Sheppard immediately with his umbrella. The umbrella dropped to the pavement, and the two men exchanged punches. Sheppard fell, and a friend who had accompanied him yelled: "Don't strike him, Captain FitzRoy, now he's down!" And FitzRoy stepped back.

Letters to the press followed, from each man expressing his outrage and attempting to justify his action. Colonel Pringle Taylor, who had been Sheppard's second, wrote a letter referring to Captain FitzRoy's "lost station as a gentleman." Both men published sixpenny pamphlets explaining their versions of the affair: *The Conduct of Captain Robert FitzRoy R.N. in reference to the Electors of Durham and the Laws of Honour, exposed by William Sheppard Esquire*; and *Captain FitzRoy's Statement re Collision between William Sheppard and the Author*.

The episode was small beer, except to Sheppard and FitzRoy. It was soon forgotten, and FitzRoy entered into his life as a member of Parliament. He was competent. He spoke on colonial matters, naval armament, and the building of breakwaters. He took an interest in matters that tapped into his ardent Christian wish to uplift those less fortunate than himself—earlier acted out on the Fuegians—such as the provisions of the Poor Law, which attempted to provide for the destitute. He proposed a bill to require and regulate the examination of seamen wishing to become masters or chief mates aboard merchant vessels, making certificates, or masters and mates "tickets" as they are known today, compulsory (eventually made law by the Mercantile Marine Act of 1850). He was made an Elder of Trinity House, the organization responsible for the building and maintenance of Britain's lighthouses, buoys, and aids to navigation. He was appointed Acting Conservator of the River Mersey. But FitzRoy didn't enter Parliament to pass his days in sleepy if honorable legislation. He saw it as a stepping stone to some greater glory, something that would use his talents and experience, rather than his lineage. The apparently tailor-made opportunity appeared just a year after his election.

In 1842, Archduke Frederick of Austria brought his warship to Plymouth for repairs, and FitzRoy, still a naval officer, was directed by the Admiralty to accompany the Archduke on a two-and-a-half-month-long tour of England, which included naval inspections of Woolwich Dockyard and Arsenal. The *Times* covered this visit, and within a few months FitzRoy's profile grew statesmanlike. So when Royal Navy Captain Hobson, the first Lieutenant Governor of New Zealand, died of a stroke that same year, FitzRoy suddenly appeared an obvious replacement. He had been to New Zealand—no ordinary thing in 1842. He had taken an active interest in colonial affairs in Parliament. It was felt by some that he had never been properly rewarded for his services to the Crown as a navigator and surveyor. The secretary of the Church Missionary Society, who had close ties to the Colonial Office, was still Dandeson Coates, who in 1830 had helped FitzRoy to secure a place for his three Fuegians at Walthamstow and was later the primary contributor of the goods (those tea services and beaver hats) they had taken back to Tierra del Fuego with them. Coates's approval of FitzRoy carried weight in the Colonial Office. FitzRoy's experience with difficult natives was well known and exemplary. And finally, his class and aristocratic background were most suitable. He was appointed the new governor of New Zealand, ending his career as a member of Parliament.

The main "situation" to be dealt with in New Zealand was the perennial constant of every nineteenth-century colony: the relationship between the white colonizers and the natives. New Zealand's Maoris were confirmed cannibals and ferocious fighters, and foreign powers had been hesitant to take them on, but the abundant sealing and whaling grounds around the two islands had made them a desirable possession. The arrival of a French Catholic mission in New Zealand at nearly the same time as a wave of British immigrants had finally prompted the British government to act, securing the country for itself. It declared that the Australian colony of New South Wales was extended to include as much of New Zealand as could be purchased from the Maoris. The government didn't want to subdue the natives—apart from any other reason, it didn't have the troops—and this ruling recognized the independence of the Maoris. Captain Hobson had been sent to persuade them to accept British sovereignty. His main inducement was to offer them the laws and protection of the white government against the lawlessness and land theft of the whites. Matters were naturally delicate. To protect the Maoris—who had been given "all the rights and privileges of British subjects"—Hobson appointed a land commissioner to review the agreements of all land purchases made between Maoris and the immigrants and their agents. As shiploads of British settlers began arriving in Cook's Strait, they faced delays taking possession of their properties in the new settlements of Wellington, Nelson, and New Plymouth. The local government was caught between the inevitably conflicting interests of the natives and the newcomers. Relations between whites, natives, and government had already deteriorated by the time of Hobson's death.

FitzRoy, his wife, and now three small children, sailed from England in July 1843. This voyage—down the Atlantic, around the Cape of Good Hope, and eastward halfway across the world skirting the northern edge of the Roaring Forties—was hardly yet routine or safe. It was five and a half months before they reached Auckland, where Hobson had set up the government's seat of office. While they were at sea, a property dispute over land in the Wairau Valley southeast of Nelson between an inexperienced magistrate, a rabble of white immigrants, and two Maori chiefs turned into a skirmish. The Maoris prevailed, most of the whites ran away, and the situation might have subsided had the Maoris not tomahawked to death their nineteen white prisoners.

The FitzRoys reached Auckland in December 1843. They stepped ashore to meet a cheerful welcoming committee from the "native department," local officials, soldiers, and a small band with two drummer boys.

"I have come among you to do all the good I can," announced FitzRoy. The band began to play, marching His Excellency and family to Government House.

Mary FitzRoy wrote home to FitzRoy's sister, Fanny Rice-Trevor in England, that they had been met by "a respectable show of ladies." The challenge of the new appointment invigorated FitzRoy. Robert, she wrote, was up every morning at 5: "he looks very well and is as cheerful as possible." The governor's house, recently built, unfinished, and suffering months of disuse, was primitive after London, but Mrs. FitzRoy did her best with it.

> We are to have two little dinner parties this week, but except in a small way there can be no society when people are so scattered and have no carriages: the nearest family live at a great distance, and except them I hear of nobody to associate with of the better kind, but among the subordinates there are some pleasing and good people. . . . The natives are certainly a most intelligent, interesting race—many very well dressed in European clothing have been with us at different meals and behaved *perfectly*. . . . They appear to understand every measure of govt. thoroughly.

Presumably, her husband's opinion of the natives agreed with her own. The Maoris were definitely a cut above the Fuegians, though both Darwin and FitzRoy had agreed in 1835, fresh from Tahiti, that they did not measure up to the Tahitians in grace or intelligence. But by 1843, many Maoris had been accustomed to living among and in tandem with whites for many years, and the English found them, on the whole, amenable neighbors. Syms Covington's 1835 description of New Zealand settlements would have described Auckland in 1843. "Here are a great many English and Church Missionaries. Chapel with organ and many well built houses besides the Native or Indian huts which are built of rushes etc."

Once the house was tidied and the two little dinner parties were over, FitzRoy sailed aboard the naval frigate *North Star* for

Nelson, to clear up the still smoldering aftermath of the Wairau massacre. He made a hash of it.

In Nelson, he spoke with the local magistrates. Then he traveled with a small group of whites—the magistrates, officials, missionaries, settlers—to a *pas*, or fortified village, to meet with the two Maori chiefs who had killed the nineteen whites. He arrived on a Sunday. One of the chiefs, Te Rauperaha, had recently converted to Christianity, and FitzRoy attended a church service at the *pas*.

The official meeting between FitzRoy and the chiefs, attended by the whites and 500 Maoris, was held the next day. FitzRoy had prepared some opening remarks:

> [On first hearing of the massacre] I was exceedingly angry; my heart was very dark, and my mind was filled with gloom. My first thought was to revenge the death of . . . the *paheka* [white men] . . . and for that purpose to bring many ships of war, sailing vessels, and vessels moved by fire, with many soldiers; and had I done so, you would have been sacrificed and your *pas* destroyed. But when I considered, I saw that the *paheka* had, in the first instance, been very much to blame; and I determined to come down and enquire into all the circumstances, and see who was really in the wrong. I . . . have heard the white man's story; now I have come here—tell me your story, the native's story, that I may judge between them.

Te Rauperaha explained that the Maoris believed the land in question had not been legally purchased. He described the arrival of the Nelson magistrate, who had waved a pair of handcuffs and threatened to handcuff the chief. Apart from the indignity this would have caused him, Te Rauperaha told FitzRoy he would rather have died than be restrained. He said the whites had opened fire, that Maori women and children—including Te Rauperaha's own daughter—had been killed. As to killing the prisoners, it was, he said, the Maori custom to kill captured chiefs.

FitzRoy questioned the chief on some points, and then was silent for half an hour, making notes, considering his judgment. Then he spoke again.

> Listen, O ye chiefs and elder men here assembled, to my words. I have now heard the Maori statement and the *paheka* statement of the Wairau affair; and I have made my decision. In the first place the white men were in the wrong. They had no right to survey the land which you had not sold until Mr Spain [the land commissioner] had finished his enquiry; they had no right to build the houses they did on that land. As they were, then, first in the wrong, I will not avenge their deaths.

He repeated the last phrase several times, in English and in translation.

> But although I will not avenge the deaths of the *pahekas* who were killed at Wairau, I have to tell you that you committed a horrible crime in murdering men who had surrendered themselves in reliance on your honour as chiefs. White men never kill their prisoners. For the future let us live peaceably and amicably—the *paheka* with the native, and the Maori with the *paheka*; and let there be no more bloodshed.

Naturally, this solemn judgment enraged the whites. Local opinion was that the new governor was weak, an imbecile, a coward. The Maoris held him in contempt for his weakness. Te Rauperaha was reported as saying that he would eat the governor and his ship. And nobody liked his pompousness.

It was an impossible situation for FitzRoy. His course of action was not unreasonable: an arrest—even if that had been possible—and trial of the two chiefs would only have exacerbated tensions. The tribes would have stepped up hostilities against the settlers; British troops, barracked in warships, would have proven ineffective at protecting them or chasing down

Maoris in the bush. More skirmishes, and outright warfare, might have resulted. It was no way to further British interests or to govern.

But FitzRoy's decision on the Wairau affair set the tone of his governorship. It was messy and inconsistent, and it pleased no one. The fledgling colony was receiving no money from England, which felt that it should support itself. With the need to produce a national revenue, FitzRoy increased customs duties, then cut them when this proved wildly unpopular. He called in extra troops from Sydney but didn't use them. His sense of fair play for the Maoris led him to reverse a decision made by the land commissioner, giving land back to the natives that had apparently been legally purchased by the New Zealand Company. Finally, the settlers around Cook Strait sent a petition to Parliament to have their governor removed. FitzRoy's effigy was burned in Nelson. Less than two years after he arrived in New Zealand, FitzRoy was recalled.

He was fired less for his unpopularity in New Zealand than for the tendencies that had previously got him into hot water with the Admiralty: he acted without instructions or against them. Like buying and hiring schooners, FitzRoy did what he thought was best and wrote home about it later. It was good officer initiative, and he might have waited six months before hearing back from England on any idea—but it wasn't good politics, and it wasn't the way things were done.

His greatest fault was keeping his government in the dark, allowing it to hear the complaints of its subject settlers without supplying explanations for those complaints.

The official Colonial Office comment on his governorship was damning.

He was surrounded by many difficulties which increased under his system of government, and was pressed by conflicting interest, which he had not the decision and firmness to adjust. Under a constant excitement and perpetual hurry he seemed neither to have had the power to adopt judicious and constant measures nor the calmness to keep the Secretary of State informed on the acts of his Government or the motives of his actions.

FitzRoy had been too fair to the Maoris, too concerned that they not be robbed, swindled, killed, and marginalized along the inevitable lines that characterized the takeover of every other colonial prize. He did not always automatically champion his constituents, whose government at home had hired him. He was too idealistic. He was too naive.

As one historian put it: "Captain FitzRoy, as Governor, had he been endowed with very great abilities, would probably, under the circumstances, have failed; but, unhappily, his qualities were such as to make his failures certain and complete." Another wrote, "FitzRoy . . . must have been the despair of his well-wishers in that he seemed not only ignorant of how to come out of the rain but undesirous of doing so."

His successor as governor, army captain George Grey, had better instincts. He attacked the Maoris on Sundays while they prayed in their new Christian churches. He captured Te Rauperaha and put him in prison. He was charming, ruthless, political, an opportunist. He earned the thanks of the settlers and the fearful respect of the Maoris, and was knighted by his government.

As he voyaged home with his family to England—a long, stormy passage east through the Roaring Forties, and through his old cruising ground the Strait of Magellan—FitzRoy must have reflected bitterly on the promise his career had shown just three years earlier. His removal as governor had left him with the certainty that his reputation had been "deeply and irreparably injured."

Despite his unquestionable accomplishments as a surveyor, sailing through Tierra del Fuego once more could only have plunged him into deeper gloom. At some point he had received a letter from his old lieutenant, and close friend, Bartholomew Sulivan, who as commander of HMS *Philomel*, was again sailing in southern South American waters, surveying and charting the Falkland Islands. Darwin had also corresponded with Sulivan, and a footnote in the revised 1845 edition of *Journal and Researches* is the only clue of what FitzRoy would also, undoubtedly, have learned.

> Captain Sulivan, who, since his voyage in the Beagle, has been employed on the survey of the Falkland Islands, heard from a sealer in (1842?) [Darwin's parenthesis], that when in the western part of the Strait of Magellan, he was astonished by a native woman coming on board, who could talk some English. Without doubt this was Fuegia Basket. She lived (I fear the term probably bears a double interpretation) some days on board.

As he passed through the Strait of Magellan, a passage of 310 miles, which might have taken a week, FitzRoy must have scanned the shore he knew so well for hours at a time, often through a telescope. He must have wondered if he would see her.

He could not have appreciated Fuegia's evolution: that the natural adaptation of her sweet empathy, together with all she had learned while in his care, now fitted her so superbly for survival along the increasingly trafficked shores of the Strait of Magellan. FitzRoy could not have understood that for Fuegia Basket, hooking on the beach was pure Darwinism.

23

Religious mania was mainstream in nineteenth-century Britain. Almost as if in anticipation of the shattering scientific suggestions that were fast approaching, the social standing of the church reached a noonday zenith by midcentury. Churchgoing and religious observances as formalized as Japanese tea ceremonies were the foundations of respectability, the bookends of daily activities.

Church and state were inseparable partners. Religion went hand-in-glove with Britain's colonizing efforts, and the astounding success of these confirmed to every Briton the moral rightness of might. Beneath the ordered decorum of it all lay an iron chauvinism. British Protestantism in the mid-Victorian era was as serious, committed, and, in its long well-funded reach, as deadly as any sect or brigade of today or a thousand years ago. It had its dogma, its bigotry, its bombast of clergical denunciations. It hungered to embrace the whole world. It had clear geopolitical designs and, for all its piety and rectitude, was as rapacious and genocidal in its furtherance of these as Cortez or Ghengis Khan.

Believers in England concerned themselves mightily with the plight of heathens abroad. They knew that unless someone saved the natives of their far-flung, pink-mapped empire, and converted them to a belief in the one true God, those unhappy souls would perish. Out of this concern was born the missionary movement, with its own fundamentalist martyrs.

Britain's missionary societies, with their arsenals of prayer-books and clothes for the naked, were the perfect, complementary trailblazers for its empire. Missionaries like David Livingstone in Africa opened up territories and continents, sending back vivid reports, descriptions of mineral wealth, maps and population numbers. Their interests exactly paralleled their government's. Even Livingstone, revered by the whole world, including his African friends and servants, knew what would follow him, and pragmatically sought the alliance of church and political and commercial interests. "I go back to Africa to try to make an open path for commerce and Christianity," he told a crowd at Cambridge University in December 1857. Britain's interests in India, Africa, Australia, New Zealand, and the Dominion of Canada (where British missionaries were in serious competition with their Jesuitical French brothers) were clear, and by midcentury its colonies around the world were well established. The missionary path to the farthest reach of every last backwater and billabong in this realm was a well-trod furrow.

No nation, however, seemed interested in Tierra del Fuego. There was nothing there anybody wanted; nothing but suicide-inducing bleakness, appalling weather, and the most abject savages. Few in England gave these souls any thought. It would take an unusually zealous, masochistic missionary to bother trying to save them.

Allen Gardiner was such a one, but it took him decades of conscientious searching to find his Calvary at the bottom of the world. A Royal Navy officer eleven years older than Robert FitzRoy, Gardiner visited Polynesia in his twenties and was moved by the missionary efforts he saw there. At age thirty-two he retired from the navy to devote himself to missionary work.

He went to Africa with his wife, who bore him five children before dying in 1834. His efforts to save the Zulu king Dingaan and his tribe were muddied by the concerns of white traders and residents and ended with the Zulus' slaughter of great numbers of Boers and their own people. Gardiner fled Africa, severely disillusioned. He tried and failed to establish a mission in New Guinea, then again in numerous places in South America, where he was opposed and driven out by the tight, well-funded club of the Catholic Church. Gardiner returned to England disheartened, but there he read of FitzRoy's Fuegians, and the captain's efforts, a decade earlier, to set up a mission at Woollya.

To Gardiner, Tierra del Fuego seemed perfect. The absence there of everything but natives who needed saving appealed to him. Missionary work there would be untainted by political designs; it would be pure God's work. In 1841 he visited the Strait of Magellan briefly and returned to England, where he founded the Patagonian Missionary Society. Its focus, the remote, virtually unknown Patagonia, was new and therefore all the more urgent.

Plea for Patagonia

Weep! weep for Patagonia!
In darkness, oh! how deep,
Her heathen children spend their days;
Ah, who can choose but weep?
The tidings of a saviour's love
Are all unheeded there,
And precious souls are perishing
In blackness of despair.

But Patagonia's isolated purity worked against it. It had nothing to offer—the same reason the Church Missionary Society, preoccupied with Africa, had felt unable to take an interest in FitzRoy's Fuegians in 1830. Patagonia lacked cachet, and Gardiner was no Livingstone. Despite raising almost no money, Gardiner voyaged out to Tierra del Fuego in 1845 and again in 1848, in hopes of establishing a mission, first with one companion, the second time with five others. Both attempts failed. On the second, Gardiner and his men landed with three boats, two wigwams and six months of supplies on Picton Island at the eastern end of the Beagle Channel. The Fuegians quickly stole everything and within a week of their arrival the missionaries boarded the ship that had brought them out and returned again to England.

Gardiner was undaunted. He saw in his failures evidence that God was ratcheting up the stakes. Gardiner was the man for such trials. The bleaker the outlook, the more fervent were his efforts. He would not be dissuaded. He passed a point of no return. Back in England, he met the Reverend George Packenham Despard, a grim, overbearing private schoolmaster from Bristol (and/or, depending on accounts, a pastor of Lenton, Nottinghamshire), who shared Gardiner's determination to bring light to the southernmost souls on Earth. Together they hatched a plan: they would buy or have built a 120-ton brigantine to act as a floating mission station in Tierra del Fuego, while building a more permanent base in the Falkland Islands. A lady in Cheltenham donated £1000. It wasn't enough to buy a ship, but Gardiner was impatient, so he bought instead two 26-foot launches and two smaller dinghies and headed back to Tierra del Fuego.

He took with him six men. In any age, there is no shortage of people willing to embark on a hazardous adventure. Columbus and Magellan filled eight ships between them for voyages into the void. One hundred and fifty years ago, the possibilities offered by missionary service were limitless and first-rate. Later, Scott and Shackleton turned away droves after filling their crews for their desperate Antarctic voyages. In 1959, 109 years after Gardiner departed for the fourth time to Tierra del Fuego, sailor H. W. Tilman, looking for crew for a voyage in an old wooden yacht to the Southern Ocean, ran this ad in the London *Times*: "Hand [man] wanted for long voyage in small boat. No pay, no prospects, not much pleasure." Tilman received more replies than he could investigate, one from as far away as Saigon.

Gardiner's pool of applicants was even bigger, made up not only of would-be adventurers but also those gripped by religious fervor. His fellow missionary-adventurers were three pious Cornish fishermen, John Pearce, John Badcock, John Bryan; a Sunday school teacher, John Maidment, who had been recommended by the YMCA; a doctor, Richard Williams; and Joseph Erwin, a carpenter who had already been to Tierra del Fuego with Gardiner and wanted badly to go back with him. Gardiner's company, Erwin said, was like "Heaven on earth."

A ship, the *Ocean Queen*, bound for San Francisco, landed the group, their boats, and supplies on Picton Island, at the eastern entrance of the Beagle Channel, in December 1850. This was just on the other side of Navarin Island from Woollya and Button Island. Gardiner hoped to make contact with the Jemmy Button he had read of in the *Narrative of the Surveying Voyages*, who had not been heard of since FitzRoy had last seen him, reduced again to savagery, in March 1834, sixteen and a half years earlier. Here, Gardiner reasoned, was a soul who had seen the uplifting way of righteousness and would surely jump at the chance to embrace it again, and in doing so would help his mission establish a beachhead.

Almost immediately, things went badly. Threatening natives forced the men off the island and into their boats. They sailed south to Lennox Island, but both dinghies were lost in rough weather. On Lennox Island, the launches were damaged. They repaired them and sailed north again into the more protected waters of the Beagle Channel, where they put into Spaniard Harbour. But here, in bad weather, both launches were wrecked beyond repair. They dragged one ashore and used it for shelter.

Fresh food was a problem: they'd forgotten to take the powder for their shotguns off the *Ocean Queen*, so they were unable to shoot fowl or guanaco. Even though three of them were fishermen, they had poor luck fishing. Then their net was torn to shreds by ice. Gardiner took all this with sublime stoicism.

Thus the Lord has seen fit to render another means abortive, and doubtless to make His power more apparent, and show that all our help is to come immediately from Him.

The Lord sent them precious little: mussels, seaweed, and infrequent dead carcasses of fish, birds, and seals that washed up on the beach. By March they were ill with scurvy. As their hunger and illness mounted, so did their ecstasy: "Ah, I am happy day and night," wrote Richard Williams, the doctor, who might, one imagines, have been only too aware of his physical condition. "Asleep or awake, hour by hour, I am happy beyond the poor compass of language to tell."

The severity of their circumstances was clear to the natives. There was nothing to yammerschooner for here. The Englishmen were left uncommonly alone.

Gardiner had sent a letter to Bartholomew Sulivan, who had taken a leave of absence from the navy to farm in the Falkland Islands, telling him of their plans. He believed that a government ship from the Falklands would be passing through the Beagle Channel monthly to collect timber and would bring them supplies. He had also been assured by a Montevideo merchant that his trading vessels would reach their vicinity now and then and look out for them. As the months went by, none of these ships appeared.

Gardiner and his men were as marooned as castaways, but they did nothing to help themselves. They made no attempt to go for help. With the wreckage of their boats, they could have fashioned some floatable contraption to cross the Beagle Channel, only two miles in width, to look for Jemmy Button, but they lacked the initiative of FitzRoy's crew, who had once made a basket out of branches, caulked it with mud, and paddled away.

There are numerous cases of shipwrecked and deserted mariners who sustained themselves for long periods in Tierra del Fuego before being rescued by passing ships. In December 1834, on the western shore of Patagonia near Cabo Tres Montes, a wilder and more unforgiving site than any cove in the Beagle Channel, FitzRoy had picked up five American sailors who had deserted their New Bedford whaling ship more than a year earlier. Their stolen whaleboat had been wrecked, and they had been pinned to the shore for fourteen months. "Yet those five men, when received on board the *Beagle*, were in better condition, as to healthy fleshiness, colour, and actual health, than any five individuals belonging to our ship," wrote FitzRoy. They had no firearms, just two hatchets and their knives. They had lived well enough on seal flesh, shellfish, and wild celery. A few days before sighting the *Beagle*, they had managed to kill nine seals. They had made fire by using a flint to strike sparks off the steel of their hatchets. They had expected nobody, least of all God, to look out for them.

But with the peculiar, wanton passivity of those who entrust themselves so entirely into God's hands, Gardiner and his ecstatic martyrs lay down on the freezing shore and awaited their fate.

> The Lord in His providence has seen fit to bring us very low [Gardiner wrote], but all is in infinite wisdom, mercy, and love. . . . The Lord is very pitiful and of tender compassion. He knows our frames. He appoints and measures all His afflictive dispensations, and when His set time is fully come, He will either remove us to His eternal and glorious kingdom, or supply our languishing bodies with food convenient for us.

In October, the merchant ship from Montevideo finally sailed into the Beagle Channel. From its decks, the wreckage of the missionaries' boats was visible on the beach. The ship's crew found three bodies scattered around the shore in various stages of decomposition, before bad weather forced the ship back out to sea. Word reached the British naval ship, HMS *Dido*, which went to investigate. The bodies of all seven men were found. Gardiner's corpse lay beside the wreckage of a boat. Inside the boat lay his diary and a letter Gardiner had managed to scribble out to Richard Williams, unaware that his fellow missionary lay farther down the beach, already dead.

> My dear Mr Williams,
>   The Lord has seen fit to call home another of our little company. Our dear departed brother [Maidment, the Sunday school teacher] left the boat on Tuesday afternoon and has not since returned. Doubtless he is in the presence of his Redeemer, whom he served faithfully . . . I neither hunger nor thirst. . . .

The public reacted in horror when the news reached England. But the *Times* expressed anger and a call for common sense.

> Neither reverance for the cause in which they were engaged nor admiration of the lofty qualities of the leader of the party, can blind our eyes to the unutterable folly of the enterprise as it was conducted, or smother the expression of natural indignation against those who could wantonly risk so many valuable lives on so hopeless an expedition. . . . Let us hear no more of Patagonian missions!

But Gardiner and his fellow martyrs had finally struck pay dirt. The story put Tierra del Fuego and the Patagonian Mission Society on the map. People suddenly wanted to hear a great deal more of Patagonian missions. "With God's help the mission shall be maintained!" the society's secretary, Reverend Despard fired back to the *Times*. And the money began to flow.

The public saw in Gardiner's woeful demise the same providential gauntlet he had seen thrown down before him. The episode begged redress from a caring nation. Even Bartholomew Sulivan saw in the deaths, and in his own glancing involvement, the workings of a divine scheme. The letter Gardiner had sent to him, advising him of the missionaries' position and hopes of resupply from the Falklands, arrived after Sulivan had left the islands. "Is it not another proof that their deaths were the appointed means for carrying on the mission?" he suggested.

Thousands thought so. Patagonia suddenly acquired the sort of cachet that Tibet later enjoyed in Hollywood. The Patagonian Mission Society swelled with recognition, patronage, and donations. Despard now determined to implement the earlier plan he and Gardiner had conceived: to buy a ship, establish a mission base in the Falkland Islands, and run the ship between the base and a suppliable foothold in Tierra del Fuego. Fuegians might then be induced to come to the Falklands base, where they could be educated with benign guidance, then be returned to their homeland to spread that civilizing influence—FitzRoy's original scheme exactly. Despard wrote to FitzRoy for his imprimatur; his reputation might have been tarnished in New Zealand, but in matters Fuegian—especially when mixed with religion—FitzRoy remained a uniquely experienced authority. The captain wrote back with lofty reserve.

> I have given the subject of your letter my best consideration. It appears to me that your present plan is practicable and comparatively safe, that it offers a fairer prospect of success than most Missionary enterprises at their commencement, and that it would be difficult to suggest one less objectionable.

That was good enough. FitzRoy's approval made it kosher.

The society steamed full ahead. A ship of 88 "registered tons" (about 120 tons of displacement) was bought and outfitted. The society's publication, *The Voice of Pity*, described its Elysian vision of a new Tierra del Fuego, an ambitious advance beyond anything Robert FitzRoy had ever dared imagine. A place of gardens, and farms and industrious villages . . . [where] the church-going bell may awaken these silent forests; and round its cheerful hearth and kind teachers, the Sunday school may assemble the now joyless children of Navarin Island. The mariner may run his battered ship into Lennox Harbour, and leave her to the care of Fuegian caulkers and carpenters; and after rambling through the streets of a thriving sea port town, he may turn aside to read the papers in the Gardiner Institution, or may step into the week-evening service in the Richard Williams chapel.

Here was the wildest Victorian pipe dream of colonization from the opium of faith and ignorance. It gave no thought to the absence of any need for such a seaport town at the bottom of the world (the Strait of Magellan *maybe*, but not in the Beagle Channel where there was, as Gardiner had proved, absolutely no traffic), or to the likelihood of befriending and training all those industrious, mysteriously motivated Fuegian caulkers and carpenters. But it was a vision born of the times and it had, at last, no lack of subscribers.

The Patagonian Missionary Society's new ship, the schooner *Allen Gardiner*, sailed from England in October 1854, bound for the Falkland Islands. Its captain was William Parker Snow, the son of a Royal Navy lieutenant who had seen action at the battle of Trafalgar. Snow had also joined the navy, but left at sixteen to pursue a remarkable knockabout career on land and sea between England, where he took dictation as Thomas Macaulay spoke his *History of England*; Australia, where he wandered the outback and ran a hotel; Africa, where he rescued a shipmate from a shark; and the Arctic, where he went on one of the many unsuccessful expeditions searching for the lost Sir John Franklin. Snow wrote a book about his action-packed life, *A Two Years' Cruise off Tierra del Fuego, the Falkland Islands, Patagonia, and in the River Plate: A Narrative of Life in the Southern Seas*. It was in good part concerned with his unhappy relationship with the Patagonian Missionary Society, which began to disintegrate from the beginning. When he took the command of the *Allen Gardiner*, he was thirty-seven years old, unusually capable, and strongly opinionated. He was a believer, but he didn't get on with his missionary passengers, in particular the zealous catechist Garland Phillips, or his equally religious and fractious crew, two of whom he discharged in Montevideo.

The ship reached the Falkland Islands in January 1855. Snow and his team of missionaries set about acquiring land and building a house on Keppel Island. They bickered and disagreed about everything. The British governor, George Rennie, found them a nuisance, and was offended by hearing that the Reverend Despard had described the Falkland inhabitants as "depraved, low, and immoral." Snow got away several times by sailing back to Montevideo in the hope of picking up the society's mission supervisor, a Mr. Verity, who had been delayed in England, but in August he learned that Verity had been arrested on a bankruptcy charge. Finally, in October, a year after leaving England, Snow, Phillips, and some of the other missionaries sailed for Tierra del Fuego.

Early in November, the schooner passed through Murray Narrows. As it neared Button Island, making for Woollya— FitzRoy's cherished spot for a mission—two canoes put out from the island's shore, filled with waving Fuegians. They approached the *Allen Gardiner*'s stern.

> Standing on the raised platform aft, I sang out to the natives . . . [wrote Snow] "Jemmy Button? Jemmy Button?" To my amazement and joy—almost rendering me for a moment speechless— an answer came from one of the four men in the [first] canoe, "Yes, yes; Jam-mes Button, Jam-mes Button!" at the same time pointing to the second canoe, which had nearly got alongside.

As the second canoe came alongside the ship, a "stout, wild and shaggy-looking man" rose from it and said, "Jam-mes Button, *me*!" and asked for the ship's ladder.

Snow ordered the crew to round the schooner up into the wind, shorten sail, and lower the ladder. In a moment, Jemmy Button, fat and naked, stood on deck.

Very possibly Jemmy had met and talked with the crews of sealing vessels—as Fuegia Basket had—in the intervening years, but Snow's meeting with him was the first recorded identification of Jemmy Button since FitzRoy had said good-bye to him in the same spot in March 1834, twenty-one and a half years earlier.

As Snow anchored his ship, at least sixty or seventy Fuegians surrounded it in canoes. Some of the men in the canoes, Jemmy told him, were "bad men"—Jemmy's eternal enemies the Oensmen, whom he usually blamed for any misfortune. Snow could see no difference between the "bad men" and the rest of the Fuegians in their canoes, but he instructed his crew to be on their guard and allowed only Jemmy's uncle, two brothers, and Jemmy's daughter's boyfriend to board the ship.

As soon as Jemmy understood that there was a woman on board (Captain Snow's wife) he asked for trousers, and when he had put them on, said

> "Want braces" as distinctly as I could utter the words. In fact he appeared suddenly to call to mind many things. His tongue was, as it were, loosened: and words, after a moment's thought, came to his memory expressive of what he wished to say. There was no connected talk from him; but broken sentences, abrupt and pithy. Short inquiries, and sometimes painful efforts to explain himself were made, with, however, an evident pleasure in being again able to converse with someone in the "Ingliss talk." That he must have been greatly attached to it, is evident from the fact, that he had not omitted to teach his wife, children, and relations. I could hardly credit my senses, when I heard Mrs Jemmy Button from the canoe calling aloud for her husband to come to her.

Snow took him below to his cabin to give him more clothes. He couldn't keep his eyes off Jemmy.

I had been amongst numbers of the Aborigines in various lands: but I had never before fallen in with one who had been transplanted to the highest fields of intellectual knowledge, and then restored to his original and barren state. It was therefore with a curious eye that I scanned this travelled Fuegian. . . . He was a rather corpulent man, with the usual broad features, and moderately dwarfish in stature, his height being about 5 feet 3 inches. My clothes I found were small for him in size: but I think if he had been properly clothed and cleaned, he would have looked not unlike a bold and sturdy man-o-war's-man. As it was, with his shaggy hair and begrimed countenance, I could not help assimilating him to some huge baboon dressed up for the occasion.

Jemmy then ate supper with Captain Snow and his wife. His power of articulation largely deserted him, but Snow pulled FitzRoy's narrative from his bookshelves and tried several of the "Tekeenica" words FitzRoy had provided in the Fuegian vocabulary of his appendix, and some of these Jemmy understood. Snow showed him FitzRoy's book and explained that a good part of it was about him.

The portraits of himself and the other Fuegians made him laugh and look sad alternately, as the two characters he was represented in, savage and civilized, came before his eye. Perhaps he was calling to mind his combed hair, washed face, and dandy dress, with the polished boots it is said he so much delighted in.

Jemmy spoke of England, "Ingliss conetree," and "Capen Fitzoy, Byno, Bennet, Wilson and Walamstow" with apparently great feeling. But when Snow asked him to return with them to the mission station in the Falklands, Jemmy steadfastly refused. This had been one of the main charges in Snow's instructions: to induce Jemmy Button and his family to spend time at the mission there. But Jemmy had had enough of sailing away with Englishmen. He wouldn't go.

Snow didn't pressure him. Perhaps his sensitivity to the extraordinary dislocations Jemmy had already experienced stayed his appeals. He had the grace to let Jemmy be. After showering the Fuegian, his wife, and family with clothes and gifts, the *Allen Gardiner* sailed back to the Falklands.

The Patagonia Missionary Society was displeased with this setback of its master plan. The Reverend George Packenham Despard, frustrated by apparent inaction and failure, decided to take the place of the felonious Mr. Verity as the mission's senior supervisor and set sail for the Falklands with his family and a shipload of furniture. He arrived in August 1856, with sixteen others, among them Allen Gardiner Jr., the son of the martyred founder of the mission.

Very soon, Despard and Snow quarreled, and Despard dismissed him on the spot. He gave the captain and his wife three hours to get off the *Allen Gardiner*, their home for more than a year. He refused to give Snow any compensation or money for the couple's passage home to England. Snow and his wife soon boarded a vessel, and by December they were back in England, where Snow immediately began suing the Patagonian Missionary Society for wrongful dismissal. He began writing his book, and also published an angry pamphlet, *The Patagonian Missionary Society, and some truths connected with it.*

But under Despard, the mission flourished. The mission house on Keppel Island was enlarged, cabins were built around it, livestock was purchased and reared; a settlement came into being. A new captain, Robert Fell, was appointed to the command of the *Allen Gardiner.* In June 1858, Despard sailed to Tierra del Fuego, determined to lure Jemmy Button—the hapless talisman and focus of so many people's earnest ambitions—back to Keppel Island. Despard, young Gardiner, and Charles Turpin, another missionary, spent a week pressuring, pestering, pursuing him around Button Island, offering him who knows what riches in this life and glories in the next, and Jemmy, once and fatefully dazzled by all things "Ingliss," was finally unable to resist them. Jemmy, the older of his two wives, and three of his children, agreed to return to the Falklands with Despard in the *Allen Gardiner*, and live there for five months of further social and religious study.

It was not a happy sabbatical for the natives. They were housed in a 10-foot-square brick hut, which was soon called Button Villa. Great emphasis was placed on their strict adherence to good manners and housekeeping: wiping their boots upon entering any house; table etiquette; floor sweeping. They were taught hymns and went to church every day. Gardiner Jr., Turpin, and the others instructed the Fuegians in English, and also did their best to learn Fuegian. They tried to translate the Lord's Prayer into Fuegian with mixed results: "Dead Father, who art in . . ." There was no Fuegian word for Heaven.

For the Buttons, the experience was much like being in prison. Fed, watered, instructed, they cleaned, worked, and prayed to a rigorous, unrelenting schedule. They were never accepted as equals at the mission; rather, they were treated and celebrated as performing monkeys. Jemmy was constantly admonished for what was seen as his chronic idleness. And there was always the suspicion of theft, thought to be endemic with Fuegians. Jemmy's wife was wrongly accused of stealing fence paling to use as firewood, and Jemmy was properly angry. They soon longed to return to their "contree."

The *Allen Gardiner* took them home in late November. The ship remained at Woollya for a month while Despard and his crew, with Fuegian help, built a small wooden house to stand as the mission's toehold on the Fuegian shore—the founding of the Elysian Fuegian harbor town envisaged by the Patagonian Missionary Society.

Jemmy and his family had proved unsatisfactory as acolytes—except for their publicity value in the *Voice of Pity*, which delighted subscribers in England who read of the natives' great progress in issue after issue. Despard wanted younger, newer blood, and with the promise of clothes and other gifts, he lured nine other Fuegians onto the ship to return to the Falklands with him.

On Keppel Island, these newcomers faced the same prison treatment. They were woken at 7 A.M. each morning, made to sweep and clean their house, wash and dress. Prayers followed. Then they were put to work around the settlement. This was mainly hard physical labor: digging peat or garden trenches, carrying stones; the women were taught to weave baskets from grasses and saplings. They were the mission's slave force and were only too aware of the caste barrier that existed between them and their "benefactors."

These nine Fuegians remained on Keppel Island for nine months. They were homesick and unhappy. They were frequently accused of stealing, and though they were often guilty, the thefts were small: a comb, some turnips, items of no great value. The missionaries' scowling, Bible-thumping reprimands infuriated the Fuegians, to whom the accusation of theft—in their society made only in the most serious of cases, such as the theft of a canoe or a wife—was a particularly black slur. Relations between the missionaries and their guests worsened and continued to the very end of their stay. As they were boarding the *Allen Gardiner* to go home in late September 1859, they were subjected to a search as thorough as that now conducted at any airport. It revealed sundry small tools, rags, bits of animal carcasses, boxes of biscuits. Not much to show for nine months of hard labor, but it was too much for Despard; it was sin and he wouldn't overlook it. The Fuegians were outraged and ripped off their clothes and threw them into the water from the gangplank. Later, they retrieved their clothes and put them on again, their feelings still outraged, but they were headed for home.

To the Fuegians' further frustration, the voyage home, which should have lasted no more than three or four days, took three weeks. The ship stopped off in the Falklands capital, Port Stanley, for five days; once it reached Tierra del Fuego, stops were made at various anchorages, gales delayed progress, and the Fuegians rightly felt the Englishmen were being insensitive to their now ardent desire to set foot on their own soil again.

The ship reached Woollya on November 2. Jemmy Button, "naked, and as wild-looking as ever," recorded Captain Fell in his diary, immediately came alongside in his canoe and boarded the ship expecting gifts, as the Fuegians' premier ambassador. What he got wasn't enough to please him, and he went away angry. Captain Fell lacked his predecessor Captain Snow's more sensitive touch. Fell also had the nine returning Fuegians searched again before leaving the ship. Two of them, Macalwense and Schwaiamugunjiz, or Squire Muggins as he was called, attacked Fell, though he pushed them off. The outraged Fuegians again flung down their blankets and boxes, tore off their white man's clothes, climbed over the rail, and paddled away in waiting canoes. Fell later brought their belongings ashore, together with more clothes and gifts for Jemmy Button. The missionaries got their Fuegians back to work cutting wood for new buildings for the settlement ashore in Woollya. But unease and ill-feeling lingered.

On Sunday, November 6, four days after the ship's arrival, the entire complement of the ship's English crew, with the exception of the cook, Alfred Coles, rowed ashore for a church service in their small wooden building. About 300 Fuegians were camped on the beach around the building. As the voices inside rose in a hymn, Coles, out on the *Allen Gardiner*, saw the natives begin to move. A group of them ran to the ship's boat and snatched the oars, carrying them away to a wigwam, and pushed the boat out into the water off the beach. The rest swarmed around the small building with clubs and spears. The doors were pushed open, and Coles heard the singing stop. He heard shouts and yells. He saw the Englishmen fight their way out of the building, to be clubbed to the ground and speared by the mob of natives. He saw August Petersen, one of the seamen,

break away from the group and rush to the water. Garland
Phillips ran after him. They splashed through the shallows after
the drifting boat. Coles saw Tommy Button, Jemmy's brother,
hurl a stone that hit Phillips in the temple, dropping him into the
water. Another stone hit Petersen. Coles watched them both
drown. He saw Captain Fell and the remainder of the Englishmen, eight of them altogether, clubbed and speared to death on
the beach.

Coles jumped into the ship's dinghy and rowed across the
harbor. He was pursued by native canoes but reached the shore
ahead of them and disappeared into the woods.

Four months later, the American ship *Nancy*, Captain William
Smyley, sailed into Woollya cove. It had been chartered by the
Reverend Despard, who had remained behind in the Falklands
and was worried about the nonappearance of the *Allen Gardiner*. The mission's ship lay derelict at anchor. The *Nancy* hove
alongside, hailing anyone aboard and getting silence for an
answer. Soon, as always with any ship, the *Nancy* was surrounded by native canoes. From one of them, Alfred Coles
climbed up the ship's side to the deck. Jemmy Button climbed up
from another canoe and went straight to the galley for food.
While Jemmy was busy eating, Coles told his story to Smyley,
who wrote it down as he spoke.

After a few days hiding out following the massacre, Coles
had been taken in by the natives who remained friendly to him;
their anger apparently dissipated. The women had looked after
him. He spent four months with them in Woollya; the natives
had even given him one of the murdered men's guns to shoot
geese with. He had gone aboard the *Allen Gardiner* a number of
times to forage for anything useful, but the ship had been
stripped of everything by the natives. He told Smyley that he
believed Jemmy Button, angry at the poor gifts brought to him
by the ship, had instigated the massacre, whipping up the ani-

mosity for the Englishmen still harbored by Squire Muggins and the others who had returned from the Falklands.

The years of capture, handouts, and humiliation, had finally brimmed over inside Jemmy Button. His admiration and real affection for Robert FitzRoy, who had raised him up to such vertiginous heights, allowed him to glimpse and touch what he could never be, and then cast him adrift to slide back down to his primordial station, had turned to the bitterest resentment and anger.

"The boys of the tribe," Coles said to Smyley, "told me that Jemmy Button and the others went on board the *Allen Gardiner* the evening of the massacre and that Jemmy slept in the captain's cabin."

24

Robert FitzRoy read of the massacre in the British press in early May 1860. In statements provided at an official inquiry in the Falkland Islands, Jemmy Button denied any part in the massacre. He blamed the "Oens-men," the bad Fuegians who had appeared at the scene of his every misfortune. He claimed he had not slept in the murdered captain's berth. The accusations against him were hearsay, nothing was proved. FitzRoy would not have believed Jemmy capable of murder, or even of inciting the massacre, but he felt deeply the disgrace of his protégé. He despaired at the unfolding destinies of Jemmy Button and Fuegia Basket.

His own destiny had also veered far off its plotted course. In the years since his return in the Beagle, he had conspicuously failed to capitalize on that triumphant success. The recognition and the advancement he had once imagined would be the proper inheritance of his undoubted gifts and accomplishments had not materialized. Meanwhile, many close to him were seeing such hopes fulfilled: his own half-brother had become governor of Australia and been knighted for his efforts. Friends, fellow naval captains of his own age, were receiving promotions and knighthoods.

Since returning from New Zealand, he had tried to put politics and his failure as a governor behind him, and to reinvigorate his maritime career, something he still believed held prospects for him. But in the Admiralty, and in the high places where favor was bestowed upon the overflowing pool of peacetime candidates for every position, men winced unhappily when FitzRoy's name was put forward. Such early promise, they said, shaking their heads, sucking in their breath. Such brilliant achievements as a surveyor and a roving scientist; what a rare seaman and navigator; what he did with the *Beagle*. . . . But the man was difficult, he acted too much off his own bat—more than that, he was "sensitive, severe, fanatical." FitzRoy's superiors clearly recognized his strengths but they didn't know what to do with him. They saw what Darwin had seen: "some part of his brain wants mending." They knew him too well: the clubby pack of lords, dukes, viscounts, and their cronies who peopled the Admiralty and the government knew FitzRoy's family as well as their own—by one distant connection or another, he was related to half of them—and they could see he had not surmounted the dark strain in his blood that had driven his uncle to madness and suicide. They feared letting him loose, giving him his head. He was a most unfortunate case; they simply wanted him to go away.

Nowhere is it recorded what FitzRoy felt when he returned to England from New Zealand, recalled, a failure, which in his eyes equaled disgrace. Nothing to spell out the anguish of a brilliant but unbalanced man in agonized limbo through what should have been his best years, the way before him lost, the terrifying sense of insubstantiality, of unstoppable freefall—all the more so after his wife Mary died in 1853. She left him with four young children. Then in 1856, his eldest daughter, "a beautiful and charming girl," according to Darwin, died. FitzRoy's fragile hold on his sanity through this period would have been a heroic daily effort.

His fortune, which was evidently all too finite an inheritance, had taken big hits during his *Beagle* years, and again in New

Zealand. He had spent willingly and unstintingly as he had thought necessary, mostly in the service of his government, with the faith that excellence of outcome would bring reimbursement, and ultimately the sort of position (for example, a governorship) that would provide its own rewards. It hadn't worked out that way. For financial as well as emotional reasons, FitzRoy needed a job.

His search for a suitable post only resulted in what were, for a man of his rank, seniority, and talents, a series of insignificant, demeaning, dead-end posts. In 1848 he was made superintendent of Woolwich Dockyard—the *Beagle*'s home port, her commissioning yard, a place he knew well, but hardly a choice move, going from governor to dockmaster. Just six months later, he was given command of a new ship. This was better, and the ship's design made it a reasonably important commission. It was the 360-ton HMS *Arrogant*, the navy's first steamship expressly designed to be driven by screw propeller, the outcome of a lengthy and disputatious association between the Admiralty and engineer Isambard Kingdom Brunel, designer of the SS *Great Britain*, the world's largest steamship when launched in 1843, and the first designed for a screw propeller. Brunel had tried to convince their hidebound lordships at the Admiralty that propellers were superior to side paddles. Eventually they came around to believing him, but denied him public recognition for his work and incurred his lasting disgust. FitzRoy, known from his earliest days as an enthusiastic embracer of the latest scientific advances wherever they might improve a ship's performance—from the canned food he shipped aboard the *Beagle* to her lightning conduction apparatus—appeared the perfect choice. But the command and his ship took him no farther than a number of trips up- and down-Channel between Woolwich and Portsmouth dockyards for sea trials and repairs. The commission was unsatisfactory and he resigned.

For just a few months in 1853, as Turkey and Russia began to fight each other and the Crimean War loomed, FitzRoy obtained a position as private secretary to Lord Hardinge (an uncle by marriage), commander-in-chief of the British Army. This was a nepotistic handout, perhaps extended with the kindest of intentions in the months of grief following the death of his wife, but it was hardly the right job. FitzRoy must have asked for and found a galling absence of opportunity with the Admiralty, which had given his one-time junior lieutenant from the *Beagle*, Bartholomew Sulivan, a warship in which Sulivan fought a successful action against the Russians, for which he too was knighted.

FitzRoy was rescued by a humbler but more suitable appointment. The Board of Trade, acting on suggestions made at a conference of maritime powers in Brussels to devote more attention to the science of meteorology, began searching for a chief weatherman. The board approached the Royal Society, England's most exclusive club of scientists—Darwin, Lyell, and FitzRoy were among its members—for a recommendation. The society named FitzRoy. He was made the government's meteorological statist, given a small office and a staff of three, and directed, vaguely, to study the weather. This was the beginning of what is today the government's Meteorological, or "Met," Office.

There was no glory in such a post; perhaps that's why FitzRoy tried to leave it in 1857, applying for the position of chief naval officer in the Board of Trade's Maritime Department. But this was given to Sulivan, and FitzRoy went back to weather.

In truth, it suited him. The position was an obscure cubbyhole in the vast office of government. No one paid him much attention. He filed annual reports. He was left to get on with the rather nebulous job description of paying attention to the weather as it pertained to the shipping interests of Britain. And in doing so, FitzRoy discovered in himself an enthusiasm and aptitude for a line of work as ardent as Darwin's pursuit of beetles.

He began by soliciting and collecting weather observations—wind force and direction, sea currents, atmospheric pressure, air and sea temperature—from captains of British merchant and naval ships. He collated this information and began plotting it visually on sea charts. He created "wind stars"—they looked like small-hubbed wheels with spokes of uneven length indicating the force and direction of prevailing winds—and stuck them, each shaped according to received information, in the middle of 10-degree squares of sea area on a chart. FitzRoy's intention was to produce such self-evident weather charts for every month of the year for British coastal waters, and eventually for all the world's oceans. The information given on such a chart would enable navigators to pilot their vessels along the most advantageous, and safest, routes for any season. It wasn't a new idea. An almost identical weather chart system was already being developed in the United States by American naval lieutenant Matthew Maury, but no such scheme was underway in Britain and FitzRoy, recognizing Maury's brilliant innovation, was eager to adapt it to British concerns.

More than most readers of daily newspapers (and he certainly did read what he once decried as "those fritterers of the mind"), FitzRoy noticed the constant stories of death and disaster faced by fishermen along Britain's stormy, tide-swept, rocky coasts. Commercial fishing today, despite every modern aid to navigation and well-funded coast guards, is still arguably the world's most dangerous industrial occupation. In the nineteenth century, the casualties were like the steady slaughter of war. FitzRoy believed many lives could be saved by more efficient weather forecasting. He was particularly intrigued by the predictive possibilities of barometer readings. "Comparisons and the judicious inferences drawn from them afford the means of foretelling wind and weather during the next following period," he wrote in a report.

But only naval and large merchant ships were equipped with barometers. Small fishing vessels certainly didn't carry them, and even ashore in seaside communities, if not on the walls of the odd natural-philosophizing parson, there were few barometers and fewer savants who knew how to read them. FitzRoy enlisted the help of the recently formed British and Scottish Meteorological Societies, the national Lifeboat Institution, and the funding of numerous philanthropic citizens to manufacture a type of barometer that quickly became known as a FitzRoy barometer. These were gothic miniweather stations, housed in tall, narrow, glassed-in frames, their reading tubes set against elaborately lettered calibrations, providing readings of temperature and humidity in addition to a reading of atmospheric pressure.*

FitzRoy sent these instruments out to fishing villages around the country, accompanied by a fifty-page instructional *Barometer Manual*. The manual became best known for its weather rhymes—ancient, time-honored doggerel passed down through generations of fishermen—which FitzRoy collected together for the first time:

A red sky in the morning is a sailor's warning;
But a red sky at night is a sailor's delight.

When rain comes before wind,
Halyards, sheets, and braces mind!
But when wind comes before rain,
Soon you may make sail again.

Several of these rhymes were specifically aimed at barometric readings.

When rise begins, after low,
Squalls expect and clear blow.

---

*I saw a FitzRoy in poor condition go for £25 at Christie's in London on October 31, 2002. Models in better condition will fetch anywhere between several hundred to several thousand dollars.

and,

Long foretold, long last;
Short notice, soon past.

The manual also contained time-proven weather lore based on simple observations.

A grey sky in the morning, fine weather; a high dawn (when the first light in the sky is some distance above the horizon), wind; a low dawn, fair weather.

A dark, gloomy blue sky is windy; but a light bright blue sky indicates fine weather.

A bright yellow sky at sunset presages wind; a pale yellow sky, wet.

Small, inky-looking clouds foretell rain; a light scud driving across heavy clouds presages wind and rain.

Soft-looking or delicate clouds foretell fine weather with moderate or light breezes; hard-edged, oily-looking clouds, wind.

Generally speaking natural, quiet, delicate tints or colours, with soft undefined forms of clouds, foretell fine weather; but gaudy colours or unusual lines, with hard definite outlines presage wind or rain.

FitzRoy was the first to disseminate and popularize such weather lore. Today's modern sailors are increasingly glued to their satellite-enabled weather-fax machines, but some still know and learn these helpful rhymes, and the connection with older followers of the sea is one of the pleasures of their use.

It was many years before the popularizing of barometers and such weather lore affected the ingrained habits of fishermen and others who lived by the sea, but a maritime disaster in the autumn of 1859 brought a sudden national imperative to FitzRoy's efforts. On the night of October 25, the *Royal Charter*, an iron sailing clipper with an auxiliary steam engine, near the end of a long passage from Melbourne to Liverpool, was blown onto the rocky coast of Anglesey, off north Wales, by a hurricane. The ship's entire complement of passengers and crew, over 400 men, women, and children, was lost. Sea disasters were not uncommon, but the ship was a large one and the number of lives lost, without a single person saved, appalled the public. It was the nineteenth-century equivalent of the first crash of a jumbo jet; the numbers, the totality of the casualties, were shocking. The hurricane's approach was something that could only have been detected by barometer readings, and while hurricanes move with a suddenness that may have made any warning too late for the *Royal Charter*, safety from weather at sea instantly became a hot issue. It was discussed at that year's meeting of the British Association for the Advancement of Science, presided over by Queen Victoria's husband, Albert, the Prince Consort, who afterward continued discussions through two further meetings at Buckingham Palace. FitzRoy was prominent in these, and suddenly found himself well-positioned to push ahead with a new cause.

Around the coast of Britain he set up eighteen weather stations connected by telegraph to his office in London. Another six stations were established on the European coast between Portugal and Scandinavia. With the daily and hourly observations telegraphed to him, he drew charts that provided a visual synopsis of the existing and *predicted* weather for the sea areas around each of these twenty-four stations. FitzRoy called these "synoptic charts," and so they are still called today. He was then able to transmit back to the weather stations forecasts for the next day. In tandem with this, he developed a visual warning system of cones that could be displayed in ports and harbors at the approach of bad weather. This information was useful well beyond maritime communities, and soon FitzRoy was sending his synoptic charts to the newspapers, which for the first time began printing daily weather forecasts.

This was important, historic work, though it wasn't recognized at the time. But grinding away in his office, FitzRoy was promoted to the rank of rear admiral, and enjoyed the first taste of career satisfaction in twenty years. He married again, Maria Smyth, the daughter of a distant cousin. For a time, he found a balance in life.

But beyond the walls of his shipshape home and productive little office, the world was changing. Unholy currents were abroad, and the unholiest of these was snaking toward him.

25

"I n October 1838 . . . I happened to read for amusement Malthus on *Population*," Darwin wrote in the slim autobiography he penned for his family near the end of his life.

He was referring to "An Essay on the Principle of Population" published in 1798 by Thomas Malthus, one of those natural-philosophizing English country parsons Darwin might have turned into had he never met Robert FitzRoy. He began reading it on September 28 and finished on October 3. Hardly amusing, it was a grim little monograph. Malthus had stated that unless checked by some means, human population could double, quadruple, and continue to multiply geometrically until it quickly grew beyond any possibility of feeding itself with a food supply that could only increase arithmetically and never keep up with demand. Continual global famine was the only possible mathematical result.

But Malthus observed that this didn't happen, that natural "checks"—disease, war, periodic localized famine, sexual abstinence, and early death—kept population numbers roughly at sustainable levels.

This was what Darwin, with all his cogitation, had been waiting to read. Doors opened in his brain.

It at once struck me that under these circumstances favourable variations would tend to be preserved, and unfavourable ones to be destroyed. The result of this would be the formation of new species. Here, then, I had at last got a theory by which to work; but I was so anxious to avoid prejudice, that I determined not for some time to write even the briefest sketch of it.

The theory supported his developing ideas about the transmutation of species: that adaptation to a hostile world which kept population numbers steady by attrition would, in time, produce better-adapted, survivalist species. But the implications of such a theory—godless creation, and humans from apes—were so inflammatory that Darwin resisted putting his thoughts in writing until he had the roundest possible argument to support it.

It is almost impossible today to understand the reluctance Darwin felt about publishing his ideas. While scientists, clergymen, and liberal thinkers might debate the literal or metaphorical length of the "days" of creation in the book of Genesis, the public and private acceptance of God's responsibility for it all, whether it had taken six days or six million years, was absolute. To suggest otherwise, particularly if well-supported by scientific argument, would change the way mankind perceived itself. Whether people wanted to believe it or not, Darwin's argument threatened to undermine the deepest faith. The idea of expressing it, he wrote to Joseph Hooker, an eminent botanist to whom he sounded his theory, felt "like confessing a murder."

Instead, he simply made notes. Years went by as he studied and published his findings about barnacles and other natural mysteries, while refining his thoughts about this Malthusian-style natural preservation, or survival, of the fittest individuals of any genus.

Darwin was in no hurry. Although in time his books would become best-sellers and bring him an enviable income, he felt no financial need to publish. He had a good income from his father, and in 1839 he married his cousin, Emma Wedgwood, daughter of his favorite uncle Josiah, the Wedgwood pottery tycoon who had proved so influential in his youth, particularly supporting his voyage aboard the *Beagle* when his father had disapproved. Emma brought with her real wealth, and removed forever any worry Darwin might have felt about money. This enabled him to concentrate solely on his work for its own sake, writing up his conclusions—or not—without regard to any schedule. To pursue this without the demands and distractions of London, the Darwins moved to Downe, a small village in Kent; today just on the southern edge of greater London, but then quite a rural retreat. There, with the modern equivalent of a millionaire's income and fifteen household servants, Darwin settled down to work.

He almost never went anywhere else again, except to spas for his health. Oddly, after the robustness he showed throughout his five-year voyage around the world, Darwin soon became sickly in the manner of many eminent Victorians: he suffered from chronic nausea, vomiting, headaches, indigestion, dizziness, and insomnia. The reasons remain inconclusive despite much inquiry during and after his lifetime. He may have become infected by some virus or parasite, possibly contracting Chagas' disease, during his epic travels, particularly in South America, where he ate and drank adventurously in the wild and was bitten by all sorts of insects. It may have been psychosomatic; Darwin was a great worrier about the health of his family, and his own health certainly deteriorated after the death of his ten-year-old daughter Annie in 1851. His own education and experience in medicine made him prone to a vividly imaginative hypochondria. Visits anywhere, and having guests at Downe, only made his health worse, and after his move to the country in 1842, Darwin gradually assumed the life of an invalid. He went for walks in his garden, and occasionally farther afield, but mostly he remained indoors in his cluttered study.

From here he maintained contact with the outside world through an enormous correspondence. His work depended completely on it, and perhaps no one benefited more from the remarkable efficiency of the British postal system when, during the mid-Victorian era, 25,000 postmen handled over 600 million letters a year. Newspapers, books, packages, and money orders moved around the country as fast as steam locomotives could carry them. In London there were up to eleven deliveries a day. The telegraph has been called the Victorian Internet, but its use was restricted to specialized services. It was the postal system, cheap and fantastically efficient, available to everyone, that made by far the bigger impact on people's lives and their perception of the world.

Darwin employed this incredible engine of communication to obtain data, opinions, and specimens. It acted as an enormous reference library for him. It became as crucial to his work and eventual conclusions as the great voyage that had prompted them.

Holed up in seclusion, Darwin turned away from the world and made a life of rumination and study. As he refined his ideas about species and the laws of nature, he was ineluctably led to a reappraisal of his religious beliefs.

> Whilst on board the *Beagle* I was quite orthodox, and I remember being heartily laughed at by several of the officers (though themselves orthodox) for quoting the Bible as an unanswerable authority on some point. . . . But I had gradually come . . . to see that the Old Testament from its manifestly false history of the world . . . from its attributing to God the feelings of a revengeful tyrant, was no more to be trusted than the sacred books of the Hindoos, or the beliefs of any barbarian. . . .
>
> By further reflecting that the clearest evidence would be requisite to make any sane man believe in the miracles by which Christianity is supported,—that the more we know of the fixed laws of nature the more incredible do miracles become . . . I gradually came to disbelieve in Christianity as a divine revelation.

> I was very unwilling to give up my belief . . . but I found it
> more and more difficult, with free scope given to my imagina-
> tion, to invent evidence which would suffice to convince me.
> Thus disbelief crept over me at a very slow rate, but was at last
> complete.

Darwin had at last become "an unbeliever in every thing beyond his own reason."

His eventual atheism may have helped ease his concerns over committing his ideas about species to paper. In June 1842, "I first allowed myself the satisfaction of writing a very brief abstract of my theory in pencil in 35 pages." Two years later Darwin enlarged this to 230 pages. But still he made no move to publish it—except to write a letter to Emma giving her very specific instructions for its publication in case of his death. But for the indefinite future he was in no hurry.

Years more went by while Darwin continued to make notes, and busy himself with other publications—volumes on the zoology of the *Beagle* voyage (as editor), new editions of his *Journal of Researches*, books and monographs on geology, volcanic islands, coral reefs, and barnacles. He went no further in writing out his ideas on the "species question" than in letters to friends, scientists like Lyell, and Joseph Hooker.

However, the question of how species came into existence was, in the late 1850s, gaining considerable attention. A number of naturalists and scientists were beginning to write about it, from all points of view, and Lyell knew enough about Darwin's work to advise him to go public and make this issue his own. In 1856, at Lyell's urging, Darwin finally began to write out a definitive examination of his ideas on the transmutation of species.

His treatment of it was so exhaustive that he might have gone on for a decade but for the slim package that came with the mail one day in June 1858. It was from one of Darwin's far-flung correspondents, Alfred Russel Wallace, posted from Ternate, a remote dot in the spice islands, or Moluccas, on the far side of the world. It was dated February 1858, and had been making its way to Darwin for four months.

Wallace, a thirty-five-year-old English naturalist, was halfway through eight years of wandering through the Malay Archipelago, the vast scattering of large and small islands astride the equator, across 45 degrees of longitude and three time zones, comprising roughly what is now the Republic of Indonesia. He was a very different sort of traveler than Charles Darwin. He came from impoverished circumstances and there was no one underwriting his travels. He had previously spent four years in Brazil, supporting himself by shipping specimens back to England and selling them to museums and collectors through an agent. In 1852 he lost all his own specimens, notes, and equipment when the ship carrying him back to England from Brazil caught fire and sank (the suggestion for A. S. Byatt's story, and the eventual film, *Angels and Insects*). In England Wallace published a book, *Travels on the Amazon and Rio Negro* that impressed nobody, sold poorly, and was soon remaindered. Undaunted, he went to Southeast Asia. With the help of Sir Roderick Murchison of the Royal Geographic Society he got a free ride aboard a government ship to Singapore, and then made his way to Borneo. From there he began again sending specimens back to England for money. His best-sellers were orangutan hides, but his most sought-after item was the rare and impossibly gorgeous bird of paradise.

In Wallace's travels through the jungles of Borneo, the volcanic island of Java, the Celebes and Banda Seas, and the fragrant Moluccas, he came across an unimaginable variety of plant and animal species. He grew keenly aware of the geographic and physical differences between them, and also between the natives of Asia, Malaysia, and Polynesia. He observed and plotted across a map of Southeast Asia a line dividing the Indo- and Austro-Malayan regions, on either side of which all the plants, animals, and even humans, belonged to these two distinct regions. Today this is still called the Wallace Line.

Wallace had also read Malthus, and his observations soon got him thinking along lines that history (always written by the victors) has termed Darwinian. He put some of these thoughts in a paper he sent to England that was published in the *Annals and Magazine of Natural History* in September 1855. Wallace later summarized the conclusions of that paper.

> Relying mainly on the well-known facts of geographical distribution and geological succession, I deduced from them the law, or generalisation, that "Every species has come into existence coincident both in Space and Time with a Pre-existing closely allied Species"; and I showed how many peculiarities in the affinities, the succession, and the distribution of the forms of life, were explained by this hypothesis, and that no important facts contradicted it.

Wallace's paper aroused various reactions. A number of naturalists felt this was pointless theorizing. Darwin read it and wrote a letter to Wallace telling him, "I can plainly see that we have thought much alike and to a certain extent have come to similar conclusions." Darwin also mentioned that Lyell too had enjoyed his paper. Wallace, whose naturalizing wanderlust had been directly inspired by both Lyell's *Principles of Geology* and Darwin's *Journal of Researches*, was pleased and encouraged. This was praise from the top. He continued his correspondence with Darwin and sent him a Javanese chicken and other skins.

In February 1858, Wallace was struck by malaria. Shivering and sweating with fever, he lay in his bed in a palm-thatched house in Ternate. Around him sat boxes of pinned butterflies, the skins and bones of birds and animals, his books, his glasses, a gun, and his sweat-stained clothing. The Malthusian question, "Why do some live and some die?" spun around and around in his fevered mind. How do some escape the natural checks on populations—and why? Hungry, lightheaded, but with a growing lucidity, the threads of all his thinking came together in an elegant conclusion. He rose from his soaked bed, staggered to his table and began writing.

> The answer was clearly that on the whole the best fitted live. From the effects of disease the most healthy escaped; from enemies, the strongest, the swiftest, or the most cunning; from famine, the best hunters or those with the best digestion; and so on. Then it suddenly flashed upon me that this self-acting process would necessarily improve the race, because in every generation the inferior would inevitably be killed off and the superior remain—that is, the fittest would survive.

Wallace worked on his thesis for three days, making it as clear and simple as he could. Finally he had a 4000-word essay that excited him. "The more I thought it over," he wrote much later, "the more I became convinced that I had at length found the long-sought-for law of nature that solved the problem of the origin of species."

He signed and dated it and sent it to Darwin, together with a letter asking if he would read it and, if he thought it worthwhile, forward it to Lyell, and perhaps help him arrange for its publication.

Four months later, the pages Wallace wrote in his hut trembled in Darwin's hands. Darwin was staggered. He could hardly believe what he read. "I never saw a more striking coincidence," he wrote in dismay to a friend. "If Wallace had my ms. sketch written out in 1842, he could not have made a better short abstract. Even his terms now stand as heads of my chapters."

Darwin was devastated. The work that had preoccupied him for twenty years, the theory he had thought his own, which he had delayed making public for so long, had now been neatly summed up by a nobody on the other side of the world. He didn't know what to do. He felt paralyzed, irresolute. So he sent Wallace's essay, as requested, to Lyell.

Lyell immediately wrote back insisting that Darwin get something of his own, a very short summary, into print immediately.

Darwin balked. "I shd be *extremely* glad *now* to publish a sketch of my general views in about a dozen pages or so," he wrote back. "But I cannot persuade myself that I can do so honourably." He was also suddenly distracted by the illness of his fifteen-year-old daughter, Henrietta, and his and Emma's tenth child, Charles, nineteen months old, both of whom suddenly came down with raging fevers. Darwin left his dilemma in Lyell's hands.

There were none better. After thirty years in the scientific limelight, defending his own revolutionary views and commanding respect, Lyell knew everything there was to know about intellectual turf and reputation. He consulted Joseph Hooker, who was also familiar with Darwin's work, and the two of them, eager to protect their friend's interests, came up with a seemingly fair solution. They would have extracts from Darwin's notes, and dated letters describing his ideas, read out together with Wallace's essay at the next meeting of the Linnean Society. This was the pre-eminent naturalist's society—Darwin, Lyell, and Hooker were members—the sort of august old boys club where Wallace ordinarily couldn't hope to have his work taken notice of. By this tactic, Darwin's years of study on the subject could be established, while Wallace would be offered the sort of respect and exposure he had never experienced.

The readings took place on July 1, 1858. Darwin's baby boy had just died and he did not attend. Nor, of course, did Wallace, then in New Guinea and entirely unaware of the whole business. The items were read. Darwin's claim to his ideas was established, along with Wallace's, and the world went about its business.

Darwin now threw aside his usual parsonly deliberation and began swiftly to do what he realized he should have done many years earlier. He began writing for publication.

The world did not suddenly shift on its axis as Darwin's and Wallace's papers were read at the Linnean Society. No outrage or damnation was voiced. Very little attention was paid to them.

History has, in retrospect, paid rapt attention to these documents and the moment of their portentous appearance, but on that July day they were simply scientific papers routinely read into the record in droning voices to a sleepy audience. Darwin was not yet Darwin, so to speak—merely a respected, reclusive naturalist noted for his works on travel, zoology, and barnacles; and Wallace was an obscure collector. What was to come of it all was still to come.

In fact, creationism, the other side of the species coin, had never been so widely and popularly discussed. In response to the challenges raised by cold science—mainly by the Lyellian view of geology and the worldwide proliferation of fossil finds, which suggested with increasing weight the enormous age of the earth and a creation that was ongoing—a slew of books appeared in the 1850s offering explanations and proofs of divine creation. Most notable and most radical among these was *Omphalos* by Philip Henry Gosse, published in 1857.

Gosse was a naturalist of growing renown. He had already published countless articles and more than twenty books (recently at the rate of four per year) on natural history, most of them about the seashore and coastal sea creatures. His self-illustrated book, *A Naturalist's Rambles on the Devonshire Coast* (1853) was a smash best-seller.

Gosse's *Rambles* was responsible for sending tens of thousands of Victorians to the seaside and started a craze for collecting shells and small sea creatures. As the pursuit and study of natural history developed into a nationwide obsession, Gosse's influence grew so great that it was reported that England had been "Gosse-ified." He was the David Attenborough of his time. His son Edmund Gosse, in his biography of his father, wrote of an incident on the rocky coast near Torbay, Devon, when Gosse, out fossicking, came upon a group of ladies who believed they had found a rare species of sea creature. Curious, but without identifying himself, Gosse asked if he might see their prize. When they showed it to him, he politely disagreed and told them it was something else, much more common. The ladies were indignant and informed him he was mistaken. "Gosse is our authority," they told him witheringly.

Gosse was the inventor of the aquarium. As a boy he had tried keeping sea anemones in a jug of seawater, but it had become cloudy and malodorous and the anemones died. He had since spent years gazing into coastal rock pools, sometimes by night with a candle or lantern. He learned that "animals absorb oxygen, and exhale or throw off carbonic acid gas; plants, on the contrary, absorb carbonic acid, and throw off oxygen." He made experiments to see how long he could keep captive sea creatures and plants together in artificial, glass-walled tanks. In 1854, he published *The Aquarium* (a term Gosse coined). It contained beautiful, expensively reproduced illustrations and plans for making "a marine aquarium for the Parlour or Conservatory." It was also about Gosse's further rambles along the seashore. It was rapturously reviewed.

> Those who have had the gratification of spirit-companionship with Mr Gosse in his former rambles, will rejoice to find themselves again by his side. . . . He has the art of throwing the "purple light" of life over the marble form of science . . . this volume ought to be upon the table of every intelligent sea-side visitor. (*Globe*, June 22, 1854)

He had become Jacques Cousteau as well. The book was another best-seller, and very financially rewarding for Gosse. No author could ask for more.

Reviewers did not remark on the religious passages in *The Aquarium*. Gosse framed many of his natural observations as proof of the existence of God, of his "wondrous contrivance in planning" and the "stable order of the universe." Such sentiments were shared by the vast majority of Gosse's readers and reviewers and were a natural complement of the fulsome prose style of the Victorian era.

Darwin and Gosse were well aware of each other. They keenly admired each other's work and corresponded frequently. Both were members of the Royal Society. They finally met at the Linnean Society in March 1855, where Gosse was reading a paper on sea anemones. Darwin was so enamored of Gosse's experiments with aquariums that he toyed with the idea of making one for himself.

Gosse knew little or nothing of Darwin's preoccupation with species, but he too had read Wallace's deduction in the *Annals and Magazine of Natural History*, that "Every species has come into existence coincident both in Space and Time with a Pre-existing closely allied Species." He had read Lyell, of course, as had most naturalists, amateur and professional, but the breadth and depth of Gosse's reading in geology was uncommon. He had explored the subject with a highly personal interest, and he had become disturbed by its inferential trend toward a belief in the natural, if not yet precisely described and identified, evolution of species.

Gosse and FitzRoy would have known each other as well, and met a number of times, for the same reasons: they were fellows of the same societies, frequenters of the same small scientific circles that included Lyell and Darwin—and in these, because of his reclusiveness, it was Darwin who was the odd man out. Until Gosse moved permanently to Devon in 1857, both he and FitzRoy lived in London and were highly active in intellectual affairs, regularly attending meetings, readings, gatherings at their societies and clubs, mixing and exchanging ideas with their peers. Both read widely and exhaustively on scientific matters, particularly in geology and the marine sciences. FitzRoy would have been well aware of Gosse's rising star, but his own had suffered a twenty-year-long downward trajectory, and when they met there was no reason for anyone to remark or remember such occasions.

The two men shared something else. Gosse was a fanatic, fundamentalist Christian. To him, God and nature were inseparable: "I cannot look at the Bible with one eye, and at Nature with the other. I must take them together," he wrote. He was a lay preacher to fellow fundamentalists who called themselves the Plymouth Brethren. He lived his life in the everyday expectation of the imminent second coming of Christ.

As marriage to Mary O'Brien had deepened FitzRoy's faith, Gosse was equally affected by the more fervent beliefs of his wife. Emily Gosse was a widely read writer of religious tracts. They had first met at an assembly of the Brethren, and their son, Edmund Gosse, wrote about the rigour of their religious faith in his book *Father and Son*.

> She [Emily] had formed a definite conception of the absolute, unmodified and historical veracity, in its direct and obvious sense, of every statement contained within the covers of the Bible. For her, and for my Father, nothing was symbolic, nothing allegorical or allusive in any part of the Scripture. . . . When they read [in the Book of Revelation] of seals broken and vials poured forth, of the star which was called Wormwood that fell from Heaven, and of men whose hair was as the hair of women and their teeth as the teeth of lions, they did not admit for a moment that these vivid mental pictures were of a poetic character, but they regarded them as positive statements, in guarded language, describing events which were to happen.

Gosse and his wife shared an intensely tender, loving, and physically intimate relationship. When Edmund was an infant and Gosse was off rambling at the seaside for his work while Emily stayed at home with the baby, they exchanged daily letters that were full of their longing for each other. "O my sweet beloved one, my helper, my comforter, my joy, my love," Gosse wrote Emily, "I wish I could just now throw my arms round your neck and kiss your dear mouth." And he signed his letters, "Ever your own faithful, affectionate, devoted, longing lover and husband, P.H. Gosse."

And Emily wrote back:

My love, How lonely you must feel tonight . . . I am always thinking of you . . . I long for tomorrow when I shall have a letter from you. Do not forget me for a moment and let me hear your assurance that you love me as I love you. I do not like to go to bed. I shall be so lonely. I miss having you to pray with me and to kiss me.

Perhaps because they married late (for the times)—Philip at age thirty-eight, and Emily, who was three and a half years older than he was, at forty-two—they were amazed and profoundly grateful for what they had found in each other. This only deepened when Emily gave birth at forty-three to their son. They couldn't believe what they had been given so late in life. "How very happy we are!" Emily wrote and said often. "Surely this cannot last!"

And it did not. In 1856, when they had been married eight years, Emily found a hard lump in her breast. After ten months of excruciating treatment, she was dead.

Philip's grief was fathomless. But he was not inconsolable: he knew for certain where Emily had gone. "He that believeth in me, though he were dead, yet shall he live; and whosoever liveth and believeth in me shall never die." He knew too that one day he would join Emily again, and as if to obliterate any doubt about this, he turned toward God as never before.

He wrote a slim book, *A Memorial of the Last Days on Earth of Emily Gosse*. Emily's religious writing had reached a wide audience, so there was more than catharsis to this endeavor.

Then he began writing something else. Faith was under assault by the findings of geology, and, with a sense of divine mission, Gosse set out to use his knowledge of natural history, his deep erudition, and his reputation to crush such heresy for good. The resulting book was titled *Omphalos: An Attempt to Untie the Geological Knot.* (*Omphalos* is the Greek word for navel.)

It could have been written by FitzRoy. Gosse's arguments and "proofs" were strikingly like those juggling geology and Holy Writ in the last two chapters of FitzRoy's *Beagle* narrative. Both used an identical technique: Scientific facts, displaying an impressive depth of knowledge, were marshaled and explained with the same appealing common-or-garden logic that had the Flood compared to a coat of varnish.

Gosse posed the eternal riddle: Which came first, the chicken or the egg? The embryo or the cow? Neither, he answered. All of organic nature consists of an endless circular process: Chicken to egg to chicken, seed to tree to seed, rain to ocean to cloud to rain, baby to man to baby.

> This, then, is the order of all organic nature. When once we are in any portion of the course, we find oursleves running in a circular groove, as endless as the course of a blind horse in a mill. It is evident that there is no one point in the history of any single creature, which is a legitimate beginning of existence. . . .
>
> Creation, however, solves the dilemma. . . . Creation, the sovereign fiat of Almighty Power, gives us the commencing point, which we in vain seek in nature. But what is creation? It is *the sudden bursting into a circle.* . . . The life history of every organism commenced at some point or other of its circular course.

But once in the circle of life, a nanosecond after creation, the very nature of the first egg showed an apparent history of its cyclical self stretching into the past.

> Its whole structure displays a series of developments . . . former conditions. . . . But what former conditions? . . . [Its] history was a perfect blank till the moment of creation. The past conditions or stages of existence can . . . be . . . inferred by legitimate deduction from the present . . . ; they are identically the same in every respect, except in this one, that they were *unreal.*

Those past "unreal" stages of the development of an organism that could be seen immediately after the instant of its creation, Gosse called *prochronic.* They *appeared* real, they could be readily inferred from an examination of the present nature of the organism. See this cow? Obviously it must once have been a calf. We "irresistibly" look backward in our belief of the existence of these prochronic stages, but they are illusory. They didn't exist before the moment of creation, the "sovereign act of power, an irruption into the circle."

Man, too, the very first one, came equipped with seemingly irresistible proof of existence stretching backward into uncountable time.

> What means this curious depression in the centre of the abdomen, and the corrugated knob which occupies the cavity? This is the NAVEL. The corrugation is the cicatrice left where once was attached the umbilical cord. . . . And thus the life of the individual Man before us passes, by a necessary retrogression, back to the life of another individual, from whose substance his own substance was formed.

Both Gosse and FitzRoy believed that man was created fully grown: "That man could have been first created in an infant, appears to my apprehension impossible," FitzRoy had written, "because—if an infant—who nursed, who fed, who protected him till able to subsist alone?"

Gosse agreed with this. When God created an organism, it hit the ground running, with an *apparent* history of its development, in the middle of an endless, ongoing cycle following the laws of nature and its own structure. Gosse's Adam was a man of "between 25 and 35 years" of age.

So it was with the fish and the fowl, and all the flora and fauna of the earth—the first towering redwood tree was created with its rings that gave an appearance of 500 years of history—so it was with the earth itself, the geological examination of which had revealed, like the rings inside a tree, the fossils embedded in its crust. In the world according to Gosse, all of nature was created with such "retrospective phenomena."

There was a freaky logic that came with all this. By the principles of Gosse's retrospective widget, the law of prochronism, the world might have popped into existence as a going concern at any recent moment, without anyone even suspecting it.

> Let us suppose that this present year 1857 had been the particular epoch in the projected life-history of the world, which the Creator selected as the era of its actual beginning. At his fiat it appears; but in what condition? Its actual condition at this moment: whatever is now existent would appear, precisely as it does appear. There would be cities filled with swarms of men; there would be houses half-built; castles fallen into ruins; pictures on artists' easels just sketched in; wardrobes filled with half-worn garments; ships sailing over the sea; marks of birds' footsteps in the mud; skeletons whitening the desert sands; human bodies in every stage of decay in the burial grounds. . . . These phenomena . . . are inseparable from the condition of the world at the selected moment of irruption into its history; because they constitute its condition; they make it what it is.

Creation might have happened five minutes ago, and all memory would be prochronic phenomena. Creation might really be a palimpsest upon a fake creation.

Gosse's theory appeared, for those who bought it, to reconcile the great paradox of the age: geology and the mosaic account of creation. But it made geology moot. If it was an illusion, what was the point? Gosse offered hope.

> The acceptance of the principles [of prochronism] . . . would not, in the least degree, affect the study of scientific geology. The character and order of the strata; their disruptions and displacements and injections; the successive floras and faunas; and all the other phenomena, would be *facts* still. They would still be, as now, legitimate subjects of examination and inquiry. I do not know that a single conclusion, now accepted, would need to be

given up, except that of actual chronology. And even in respect of this, it would be rather a modification than a relinquishment of what is at present held; we might still speak of the inconceivably long duration of the processes in question, provided we understand *ideal* instead of *actual* time; that the duration was projected in the mind of God, and not really existent.

In other words, all of scientific investigation was pointless, but still fun.

When it appeared in the fall of 1857, *Omphalos*, a new book from a best-selling author, was widely noticed and reviewed. The word most consistently applied to it was "ingenious": "The argument is startling. But it is so ingeniously framed. . . . This very ingenious analogy. . . . We cannot deny the merit of ingenuity." "His reasonings are very ingenious." "Mr. Gosse's argument appears to us both ingenious and important."

FitzRoy must have been one of its most receptive readers. Here was an esteemed author grounded, as he had been, in scientific learning, using that knowledge to reconcile, as he had tried to do, science with scripture.

Gosse wrote for FitzRoy.

But the overwhelming response to his ingenuity was scathing. Reviewers found the logic in the book "unanswerable," untestable, the theory "too monstrous for belief." The *Natural History Review* pronounced it full of "idle speculations, fit only to please a philosopher in his hours of relaxation, but hardly worthy of the serious attention of any earnest man, whether scientific or not."

Even believers were unhappy with Gosse's argument. In the *Review*'s April issue, a man with the Dickensian name of J. Beete Jukes, denounced his flippant handling of the "awe-inspiring mystery" of creation.

To a man of a really serious and religious turn of mind this treatment is far more repulsive than that even of . . . the Lamarckian School. Both classes of reasons appeal to our ignorance rather than our knowledge, and take upon themselves to make positive assertions upon things about which no man *knows*, perhaps no man ever *shall* or *can* know, anything whatever; but the soi-disant religious school to which Mr Gosse belongs has the additional bad taste to speak as if they, forsooth, were on the most intimate terms with the Creator.

Mr. Jukes was deploring the same presumption FitzRoy brought to his own scientifically buttressed logic and arguments about the Flood; the presumption of a man to explain God and his designs.

Gosse was dismayed.

In the course of that dismal winter [wrote his son Edmund], as the post began to bring in private letters, few and chilly, and public reviews, many and scornful, my Father looked in vain for the approval of the churches, and in vain for the acquiescence of the scientific societies, and in vain for the gratitude of those "thousands of thinking persons," which he had rashly assured himself of receiving. As his reconciliation of Scripture statements and geological deductions was welcomed nowhere . . . a gloom, cold and dismal, descended upon our morning teacups. . . . He had been the spoiled darling of the public, the constant favourite of the press, and now . . . he could not recover from the amazement at having offended everybody by an enterprise which had been undertaken in the cause of universal reconciliation.

He was most stung by the reaction of his old friend, the Reverend Charles Kingsley, author of *Westward Ho!* and *The Water Babies*, who had befriended and championed Gosse and his work before his books were well-known. *Omphalos* had "staggered and puzzled me," he wrote him. Kingsley also believed absolutely in divine creation, but Gosse's book was the first to make him actually doubt it.

Your book tends to prove this—that if we accept the fact of absolute Creation, God becomes a *Deus quidam deceptor.* . . . I cannot believe this of a God of truth, of Him who is Light and no darkness at all, of Him who formed the intellectual man after His own image, that he might understand and glory in His Father's works. . . . I cannot give up the painful and slow conclusion of five and twenty years' study of geology, and believe that God has written on the rocks one enormous and superfluous lie for all mankind. . . .

I would not for a thousand pounds put your book into my children's hands.

The trouble for Kingsley, and many others, was that Gosse had almost managed to convince him that only by adopting his law of prochronics could the biblical story of creation and geology be reconciled. And he found Gosse's theory so preposterous and silly that it threatened to remove for him the last barrier to disbelief. It left him gaping into an abyss.

Hardly anyone bought *Omphalos.* Most copies of the book were sold for waste paper.

For those whose faith in the literal word was buckling under the weight of scientific argument, or preposterous rationale, the last barrier came down on November 24, 1859, when Darwin finally published his book on the transmutation of species. It was not the immensely long work he had envisioned and begun a few years earlier at Lyell's urging, but, inspired by the clarity of Wallace's brief essay, a shorter, simpler volume (though still 502 pages long). It was titled *On the Origin of Species by Means of Natural Selection or the Preservation of Favoured Races in the Struggle for Life.*

The first edition of 1,250 copies, priced at fourteen shillings, sold out on the day of publication.

26

In late June 1860, FitzRoy traveled by train to Oxford to read a paper on British storms at the annual congress of the British Association for the Advancement of Science. This week-long gathering of Britain's scientific community offered exhibitions, lectures, and informal meetings for professional scientists, amateur enthusiasts, and interested members of the public. Many attendees brought their families, stayed at local inns, and enjoyed picnics and punting on the Thames.

FitzRoy came alone, leaving Maria at home in London with his children. He was in no mood for fun. Only six weeks earlier, he had read in the press of the massacre, apparently instigated by Jemmy Button, at Woollya. Everything to do with Tierra del Fuego had turned sour on him. Almost as awful as the massacre, in FitzRoy's view, was Darwin's new book. In the six months since its publication, *Origin of Species* had become a sensation.

Darwin had sent a copy to FitzRoy, in recognition of their connection and the fact that, but for FitzRoy, there would have been no book—no Darwin as history was just beginning to perceive him. This was clear to both men, horribly so to FitzRoy. He had provided Darwin with the vehicle for his conclusions. The voyage aboard the *Beagle* was the central engine behind everything his one-time friend, "Dear Philos," had accomplished since returning to England twenty-three years before.

FitzRoy hated the book. "My dear old friend," he wrote to Darwin. "I, at least, *cannot* find anything 'ennobling' in the thought of being a descendent of even the *most* ancient *Ape*." Darwin had scrupulously avoided any mention of the man-from-ape connection in his book, but it was the implication seized upon by everybody. It was the image that sold the theory.

Origin of Species left FitzRoy so disturbed that he seized any opportunity to denounce its findings or its author. When, a few days after its publication, the antiquarian Sir John Evans wrote in the *Times* about 14,000-year-old hand axes made by people of paleolithic "drift" cultures that had been found on the banks of the Somme, FitzRoy fired back a letter to the *Times* castigating Evans's conclusions. The axes, he said, were not 14,000 years old, but were left by far more recent savages who had wandered away from and lost their own civilization—the same argument he had made about Noah's wandering descendants. "In what difficulties do not those involve themselves who contend for a far greater antiquity of mankind than the learned and wise have derived from Scripture and the best tradition!" FitzRoy signed the letter not with his own name but with the pseudonym Senex ("old man"). Evans and Senex had a brief, fractious correspondence in the *Times*, during which Senex referred to "Mr Darwin" as a corroborator of Evans's "weak cause." Darwin read this and realized immediately the identity of Senex. "It is a pity he did not add his theory of the extinction of the Mastodon etc from the door of the Ark being made too small," Darwin wrote to Lyell.

It was not only fundamentalist believers who disagreed with Darwin. Many naturalists and geologists—including close friends like Charles Lyell—never fully accepted his theory of a mutating evolution ungoverned by a creator. But Lyell and others recognized the importance of Darwin's work and found their own ways to accommodate it. Charles Kingsley also received a copy of *Origin of Species*. "It awes me," he wrote Darwin, "both with the heap of facts, & the prestige of your name, & also with the clear intuition, that if you be right, I must give up much of what I have believed & written."

Yet Kingsley didn't fully comprehend the totality of Darwin's departure from holy doctrine. He still believed that no matter what clever device allowed for the evolution of species, it had all been designed and instigated by the creator. But Darwin had shown that species could evolve by an automatic mechanism that could run—and had run since the earth had cooled—independent of any designer. If Charles Kingsley failed to, many more had got Darwin's clear atheistic message: there was no creator.

What gave *Origin* its legs and brought ridicule to *Omphalos*—what brought a desperate, defensive stridency to FitzRoy's and others' protests—was that Darwin's idea of godless creation had become thinkable. By midcentury, God had suffered a decline in prestige exactly like that of the British royal family in the present era. He might exist, but he was increasingly unnecessary.

Man had become stupendously powerful. He was competing with, and exceeding, God's works. Sir Isaac Newton's handy determination of the biblical cubit as 20½ inches had revealed Noah's Ark to be 537 feet long and weighing 18,231 tons, dimensions once sufficient to contain all life on Earth. But in the 1850s there rose at the edge of the River Thames in east London a ship bigger than God's. It was the *Great Eastern,* designed by Isambard Kingdom Brunel, the little man who dreamed big in an age of mighty works. Begun in 1854, launched in 1858, the ship was 693 feet long and displaced 22,000 tons. She was the most prodigious vessel the world had ever known, the largest movable object on the planet (and almost too large to launch; it took many months and hundreds of thousands of pounds to get her across a few hundred feet of mud into the Thames). She was designed to carry 4,000 passengers, and 15,000 tons of coal, to steam without refueling around the globe to Trincomalee, Ceylon, and back to London, to monopolize, in a single voyage, the

Oriental tea trade. She weighed more, when fully laden, than the combined tonnage of the 197 English oaken ships that sailed to meet the Spanish Armada. She was vaster than the apocalyptic Rhine timber rafts that had come on the spring floods out of Germany in the Dark Ages, demolishing the towns, bridges, boats, and people in their path. Sitting zeppelin-like on the flat boggy ground on the Isle of Dogs, swarmed over by men like ants on a dead buffalo, the *Great Eastern* appeared preternaturally large, monstrous, a vision from the Book of Revelation, the fallen Wormwood star. The sight raised the hair on the napes of observers' necks. (The *Great Eastern* was in fact too large for her times and never fulfilled her dreamt-of potential; she found her most useful work laying telegraph cable across the Atlantic and was broken up for scrap in 1888.)

But nothing appeared more ungodly than the ubiquitous and universally enjoyed scientific marvel of the age, the steam locomotive. The engine itself was plainly demonic: "like a huge monster in mortal agony, whose entrails are like burning coals"; it flew "faster than fairies, faster than witches." Thomas Carlyle's hallucinatory description of a train journey at night read like a scene from Harry Potter.

> The whirl through the confused darkness on those steam wings . . . hissing and dashing on, one knew not whither. We saw the gleam of towns in the distance—unknown towns. We went over the tops of houses . . . chimney heads vainly stretching up towards us—*under* the stars; not under the clouds but among them . . . snorting, roaring we flew: likest thing to a Faust's flight on the Devil's mantle; or as if some huge steam night-bird had flung you on its back, and was sweeping through unknown space with you.

The devil, if he traveled, would go by train, said Lord Shaftesbury after journeying by rail from Manchester to Liverpool. Engineers named their locomotives *Wildfire, Dragon, Centaur.*

Rocking in half-open carriages, passengers smelled cinder-fire smoke, were deafened by screeching whistles and the snorting roar ahead, and plunged into tunnels that appeared like portals into hell. This was modern travel. There was nothing in it of the loveliness of the sea or ships, the contemplative pleasure of a walk, or the creature-communion of a ride on horseback or in a carriage.

Manmade, the railway destroyed both men and nature. Laborers cut unsightly gashes hundreds of feet into the earth for railway cuttings (delighting geologists for exposing strata and fossils as they dug). They tunneled two miles through rock at a cost of £6.25 million and 100 men's lives to dig the "monstrous and extraordinary" Great Western Railway tunnel at Box in Wiltshire (Brunel's scheme again). Collisions and derailments left horrific casualties. And yet the man-destroying science grew like a contagion, spreading its black web of track over the landscape, and more and more passengers crowded aboard trains that went faster and faster.

That speed made the remotest extremities of Britain seem infinitely closer. The Midland counties were "a mere suburb" of London. But the new speed of rail travel, like the later acceleration of information and communication, brought with it a paradoxical effect of time: there seemed to be less and less of it. People started running to catch trains. They grew anxious to be "on time." Life speeded up.

Railways captured the Victorian imagination like nothing else. Architects designed stations as vaulting iron cathedrals of the industrial age. Artists turned out endless paintings and drawings of the public crowding into stations, of railway views, of cuttings and tunnel entrances. They were also quick to see in the railway apocalyptic visions, from J. M. W. Turner's *Rain, Steam, and Speed—the Great Western Railway* (1845), in which a smoking train charging across a bridge bears down on a fleeing hare in its path, to John Martin's *The Last Judgement* (1853), showing angels gathering over a train as it hurtles into a black abyss. These said what people already knew: there would be no turning back, no matter what the cost.

Darwin's mechanistic view only coincided with such advances, but it resonated with the age; life had acquired the context in which it could be accepted. The theory of evolution now spread over the earth as a seed of truth, resembling nothing so much as, ironically, the early spread of Christianity, though incredibly accelerated. It marked the moment when one world, with all its precious assumptions and truths, was destroyed; and another, new, beguiling, and frightening, began.

As FitzRoy's train snorted through the Chiltern Hills toward Oxford (traveling at sustained speeds of 45 to 50 miles per hour, sometimes hitting 60), he could have no doubt that Darwin's book and theory would be on everybody's lips. It would be *the* topic of conversation and interest. He would hear no end of it. It depressed him terribly. His presentation of his own work, which he believed was important and was saving lives, would be a sleepy sideshow.

FitzRoy read his paper on Friday, June 29. He described some of the terrible storms that had caused loss of life around the British coast over the previous hundred years. He reviewed the disaster of the *Royal Charter*. He outlined the forecasting powers of barometric readings and described how his Met office was engaged in receiving information and providing forecasts by telegraph. And he mentioned that this work had sufficiently impressed French meteorologists to initiate a similar program.

For whatever reason—maybe he met friends and acquaintances, or had already arranged to meet them; or maybe he wanted to hear what everyone else had come to hear—FitzRoy did not return to London that night. He stayed in Oxford. The next day he made his way to the lecture hall in the University Museum of Natural History. It was the location for a lecture to be delivered by John William Draper, a chemist and historian, a Liverpudlian by birth and now head of the medical school of the City University of New York. Draper liked to mix things up. The title of the lecture he intended giving that Saturday was: "On the Intellectual Development of Europe . . ." and then came the kicker, ". . . Considered with Reference to the Views of Mr. Darwin."

Darwin had not come to Oxford. He was unwell again, or perhaps the prospect of coming and speaking in public and defending his now spectacularly controversial ideas had made him unwell. There was no lecture scheduled that was specifically concerned with "Darwinism" (the term was spontaneously coined in hundreds of conversations in late 1859–1860, and was in common use by 1861), so Draper's reference to "the Views of Mr. Darwin" singled his lecture out as the forum for the debate everyone wanted. Proponents for and against Darwin's argument, and those who wanted to hear them argue, had come to Oxford for just this lecture, and rumor had it that Samuel Wilberforce, the Bishop of Oxford, who was to be present, would use the occasion to publically denounce *Origin of Species*.

When FitzRoy reached the lecture hall, he found an overflow crowd of scientists, Oxford and Cambridge professors, journalists, and a noisy mob of undergraduates. Organizers soon moved the lecture to a much larger room in the building. As many as a thousand people pushed in, grabbed seats, stood where they could.

In an atmosphere of charged expectation, Draper started talking. Known for his dislike of organized religion, he began, promisingly, by alluding to the "Views" of his speech's title by saying that human progress was only possible when science pushed theology aside. But his listeners, ready to hear brilliance, were disappointed. Draper waffled on for an hour and a half about the intellect of the ancient Greeks, "flatulent stuff" according to Joseph Hooker, who was among the audience.

When Draper finally sat down, John Henslow, Darwin's old professor from Cambridge, who was chairing the lecture, asked if any members of the public cared to say anything. This was the

moment when everyone expected Wilberforce to rise and con-
demn Darwin, but instead a Mr. Dingle stood and said, "Let
point A be the man, and point B the monkey." Mr. Dingle had a
curious accent, pronouncing monkey "mawnkey."

"Mawnkey! Mawnkey!" undergraduates began shouting,
until Mr. Dingle could not go on.

Now Bishop Wilberforce rose, resplendent in his robes of office.
A large, self-assertive man, supposedly the model for Trollope's
Archdeacon Grantly of the Barchester chronicles. Silence was
restored to the hall. Wilberforce had written a damning but still
unpublished review of *Origin of Species*, and he used this as the
basis of his thirty-minute speech.

Darwin's book was filled with error! he boomed. His "facts"
were assumptions, not evidence, and did not support his conclusions.

"Has any one such instance [of the existence of a species by natural selection] ever been discovered? We fearlessly assert not one."

An experienced, even theatrical public speaker, deliverer of
sermons, debater, essayist and writer, Wilberforce took command of his audience. He finished on a note of withering humor:
"Is it credible that a turnip strives to become a man?"

As the room filled with laughter, Wilberforce turned and
stared at Thomas Huxley, a thirty-five-year-old zoologist who, in
six months, had become Darwin's greatest champion in print,
reviewing *Origin of Species* in a number of influential magazines
and journals. His efforts had earned him the nickname "Darwin's bulldog." Now Wilberforce decided to bait him. Was it on
his grandfather's or his grandmother's side, Wilberforce asked
Huxley, that he was descended from an ape?

While the audience laughed again, Huxley is said to have
whispered to a friend: "The Lord hath delivered him into mine
hands." He rose and responded in measured, factual terms. But
all that is remembered is his final remark, his answer to Wilberforce about his ancestry.

[As to whether] I would rather have a miserable ape for a grand-father or a man highly endowed by nature and possessed of great means and influence, and yet who employs those faculties for the mere purpose of introducing ridicule into a grave scientific discussion—I unhesitatingly affirm my preference for the ape.

Almost instantly, the meeting became legend. Nobody took down verbatim what was said, yet immediately afterward everyone who had been there began disseminating versions of the great slugfest between Wilberforce and Huxley, and opinions of who had won. The essence of all versions eventually became the same: it was the great Victorian confrontation between science and religion, between God and ape; the moment when that ceased to be the concern only of philosophers, scientists, academics, and clerics, but passed into the public consciousness and became the question of the age. History has for once determined the victors, rather than the other way around, but the battle raged furiously on. Sixty-five years later, in the famous "Monkey Trial," Tennessee schoolteacher John Thomas Scopes was convicted and fined $100 for teaching Darwinian evolution in a high school biology course, and the debate between creationists and evolutionists continues today.

Draper, whose droning lecture was the crucible for debate, is now forgotten, a historical footnote.

Also forgotten is the man who rose to his feet and tried to speak above the commotion that filled the lecture hall when Huxley was finished. A week short of his fifty-fifth birthday, he looked at least a decade older, his face pale from lack of sun, creased by anxiety and the never-ending strain of trying to find a balance between waves of upheaval, forces he had been battling all his life. He wore the naval uniform of a rear admiral.

He stood and waved a Bible over his head. Several people later wrote down what they remembered him saying: He regretted the publication of Mr. Darwin's book; Mr. Huxley's statement that it was a logical arrangement of facts was mistaken; he had often expostulated with his old friend aboard the *Beagle* for his ideas that were contradictory to the first chapter of Genesis. . . .

Few paid attention. Those who might have recognized Robert FitzRoy would have been embarrassed for him. His comments were irrelevant to either side. The roiling tide of debate swept away from him. He cut a sad figure that invited ridicule, if not pity.

He left the hall and made his way to the train station. FitzRoy felt old, passed by. Life around him appeared (apart from the late spread of the railway) much the same as it always had: hansom cabs still clip-clopped down Piccadilly, men still slumbered in their clubs, church congregations were fuller than ever, tweedy hunting parties still flushed game and shot it with fine guns, clipper ships labored their way to Ceylon and home again full of tea, and fashions changed only in small and sensible increments (hoops of baleen were replacing layers of crinoline beneath ladies' dresses, which made life more pleasant for everyone). The Great Exhibition of 1851 was a testament to Britain's indisputable dominance over the earth. But FitzRoy felt a dark shift in the world. He was a scientist; all his life he had believed in science. To him, as once for Darwin, it revealed in every aspect of its design the sure indication of an all-knowing creator. But now science was marching in directions he could no longer follow, and it left him unmoored.

He went back to his weather work. In 1862, Longmans published his *Weather Book*, 440 pages of FitzRoy's ideas and conclusions about meteorology, and his recurring mantra: "It should always be remembered that the state of the air *foretells coming* weather, rather than indicates weather that is *present*." It contained a reference to the failure of "a very young commander" to read the warning signs of the approach of a *pampero* storm off the coast of South America, which resulted in the loss overboard of two men. Also an account of "the writer's" actions as a passenger aboard a vessel on a "very tediously slow" passage to

England from New Zealand in 1846 (his disgraced trip home after being recalled as governor), when, becalmed off a "savage" shore, FitzRoy's own barometers told him of an approaching storm. The ship's captain didn't believe him, so FitzRoy, with the help of several officers, let go a second anchor, veered cable, and waited. A few hours later, violent winds came abruptly out of the night. "Had that ship been taken unprepared, not a soul would have been saved." A second edition of the *Weather Book* was published in 1863, and Longmans' records show that FitzRoy was paid £200 for the copyright.

But he could not foretell the change of the wind that now turned against him. Once the novelty of his daily forecasts passed, the mistakes in his "prophesies" were thrown back at him. His efforts began to be ridiculed. Even the *Times*, which ran his daily forecasts, poked fun at him.

> The public has not failed to notice, with interest, and, as we much fear, with some wicked amusement, that we now undertake every morning to prophesy the weather for the two next days to come. While disclaiming all credit for the occasional success, we must however demand to be held free of any responsibility for the too common failures which attend these prognostications. During the last week Nature seems to have taken special pleasure in confounding the conjectures of science.

He suffered these criticisms and jokes while continuing to do his best, but on June 18, 1864, the *Times* discontinued his forecasts. On the day of his last published forecast, the newspaper attacked its own sacked contributor:

> "Whatever," says [a correspondent], "may be the progress of the sciences, never will observers who are trustworthy and careful of their reputations venture to foretell the state of the weather." Admiral FitzRoy has still to convince the public, and at his task he labours yearly with the most praiseworthy

assiduity! . . . There can be no doubt that when Admiral
FitzRoy telegraphs, something or other is pretty sure to happen.

And with that malicious ridicule, the *Times* fired him.

FitzRoy went home and fell into "a severe attack of prostration of strength, threatening paralysis," his wife Maria wrote to a friend. She began consulting doctors "with regard to the soundness of his poor mind."

Soon afterward, they moved for his health from Onslow Square in South Kensington to Upper Norwood, just south of London, between Lambeth and Croyden, on the edge of the North Downs: only nine and a half miles from Darwin at Downe. FitzRoy continued working intermittently, as his health permitted, at the Meteorological Office, traveling into town by train.

His handwriting, in letters to friends, now grew very large.

In the spring of 1865, his condition deteriorated. Maria FitzRoy wrote an account of this time.

Friday, April 21st, in bed all day very ill . . . Sunday—up, weak and ill, unable to go out, but came downstairs, asked for his bible and prayer book, while we went to church. . . . Monday 24th, still very weak, able to take a short drive with Ad'l Cary which did him good, and then sat out with me in the garden while the girls played at croquet.

Tuesday 25th. Much better. Would go to London—did no business at his office, came back in the afternoon early. . . . Evening played at whist which he seemed to like . . .

Thursday 27th, a hot day, he started directly after breakfast for London . . . came back just after 12 very tired out, and lay down to rest, came down to luncheon. . . . In the evening he seemed quiet and happy, talking tranquilly with me alone, and seemed to have made up his mind to stay here quietly and really take care of himself. Just before going to bed he received a letter from Mr Tremlett inviting him and myself to come and stay with him from Saturday till Monday to see the last of Capt. Maury [FitzRoy's American naval counterpart at developing weather charts]. This note seemed completely to upset him, between desire to comply with his request and his just expressed wish of remaining quiet. Of course he did not sleep well that night; the only advice I gave him was to do that which would give his mind the greatest ease.

On Friday morning he went to London to his office, came back again relieved at having written and refused the invitation, so he told me. And after luncheon he went to his room to write and called me urgently to come to him. . . . I found him extremely distressed at the quantity of unanswered notes and invitations to public dinner which ought to have been answered long ago. I comforted him and helped him to answer two or three most pressing.

Saturday morning after breakfast he came to me saying he had got a strong desire to see Maury again; I told him he had better gratify it if he had; he said he was totally incapable of exertion and could only lie down and rest and asked me to make him comfortable, which I did. After luncheon he felt somewhat better, and set out to take a walk with the two eldest girls while I went for a drive. . . . When I came home I found that he had left them, and gone to London, and did not return till nearly 8 o'clock, worn out by fatigue and excitement and in a worse state of nervous restlessness than I had seen him since we left London [for Upper Norwood]. He seemed totally unable to collect his ideas or thoughts, or give any coherent answer, or make any coherent remark. After dinner he recovered a little. . . .

I was in bed when he came to bed; he came round to the side where I was, asked me if I was comfortable, kissed me, wished me good night, and then got into bed. It was just 12 o'clock. I was soon asleep. . . .

. . . in the morning I said I hoped he had slept better, as he had been so very quiet. He said he had slept he believed, but not refreshingly. . . . Just then it struck six. From 6 to 7 neither of us spoke, being both half asleep I believe. Soon after the clock struck 7 he asked if the maid was not late in calling us. I said it was Sunday, and she generally was later, as there was no hurry for breakfast on account of the train at 10 o'clock as there was on other days. The maid called us at ½ past 7. He got out of bed before I did, I can't tell exactly what time, but it must have been about ¼ to eight. He got up before I did and went to his dressing room kissing Laura as he passed through her little room.

In his dressing room a little before eight o'clock that Sunday morning, April 30, 1865, history—on his mother's side—and Darwinism, finally caught up with him. FitzRoy picked up his razor and cut his throat.

EPILOGUE

Late in 1863, two years before FitzRoy's death, Reverend Whait Stirling, the new superintendent of the Falkland Islands mission, sailed in the mission's ship *Allen Gardiner* to Tierra del Fuego. Since the massacre in 1859, relations had cooled between the missionaries and the natives, who remained fearful of some official action or punishment. But this never came, and the missionaries' hopes remained intact. They went looking for a new crop of young Fuegians to convert.

Thomas Bridges, the adopted son of Reverend Despard, creator of the Yamana-English dictionary, accompanied Stirling on that trip, acting as his translator. He learned from the Fuegians that Jemmy Button had died that year in an epidemic that raged through Tierra del Fuego.

The missionaries and the natives reestablished a connection, and eventually much friendlier relations. Over the next few years, at least fifty Yamanas visited Keppel Island in the Falklands. Stirling took four of them, including one of Jemmy Button's sons, Threeboys, to England. There, two of them achieved the sort of independence unimagined by FitzRoy's protégés, traveling around Britain by railway, sometimes alone, to speak at church meetings. They remained in England a year, a successful public relations visit for the missionary society, but Threeboys and another Fuegian boy, Uroopa, both died of illness on their return voyage home.

In 1867, Reverend Stirling and Thomas Bridges established, at last, a mission beachhead in Tierra del Fuego, at Ushuaia on the north shore of the Beagle Channel, opposite the Murray Narrows. There, among a group of visiting natives in 1873, Bridges met Fuegia Basket, then in her midfifties. Bridges (his son Lucas recorded in *Uttermost Part of the Earth*) found her "short, thickset and with many teeth missing from a mouth that was large even for a Fuegian." She evidently retained some of her once considerable charm, for she was accompanied by her current husband, an eighteen-year-old boy. She remembered some English: "knife," "fork," "little boy, little gal," but not enough for conversation. Bridges talked with her in Yamana and heard that her first husband, York Minster, had long ago been killed in retaliation for his murder of another man.

Ten years later, in February 1883, Bridges met her again, for the last time, in the western part of Tierra del Fuego. Fuegia was "nearing her end . . . in a very weak condition and an unhappy state of mind." Her young husband was gone, but "she had two brothers with children of their own [and] would lack for nothing that their circumstances could provide." Bridges tried to comfort her with "the beautiful Biblical promises in which he himself so firmly believed."

Over the years, the mission settlement at Ushuaia grew steadily from a prefabricated hut into a number of buildings; but at the same time the Fuegians were being decimated by disease to near extinction. By 1888, epidemics of measles, pneumonia, and tuberculosis had wiped out every native within thirty miles of the mission. But the remote "civilization" at Ushuaia appealed to the Argentine government, which found its desolation a perfect site for a prison. (Today Ushuaia is a rapidly expanding tourist city at the edge of Argentina's Parque Nacional Tierra del Fuego, a base for trekkers and charter yachts.)

Thomas Bridges abandoned the mission and took up sheep

ranching 40 miles east of the town. Other Europeans arrived, with more sheep. Vast farms spread across Tierra del Fuego, and the Fuegians inexorably went the way of Indians in North America: they became a nuisance, they were driven out, they were hunted down and killed. By 1910 there were fewer than 1,500 natives in all of Tierra del Fuego. Population numbers have only decreased since. Today, the Fuegians are virtually extinct.

Under new commanders, HMS *Beagle* **made two more long** surveying voyages to Australia and Southeast Asia. After returning to home waters, she was sold in 1845 to the coast guard, stripped of her name, and fitted out as a watch vessel. Her upper masts were removed and a "caboose" for the watchkeepers was built on her deck. She was moored on the Crouch and Roach rivers in Essex for twenty-five years.

In 1870, she was sold at public auction for £525. She was either broken up then or shortly after. One source has her bought by the Japanese navy and used as a training ship until 1881, and broken up in 1888.

Rear Admiral Robert FitzRoy was remembered and lamented by those who knew best what he had done. The naval hydrographer at the time of his death, Admiral George Henry Richards, was clear about FitzRoy's contribution.

> No naval officer ever did more for the practical benefit of navigation and commerce than he did, and did it too with a means and at an expense to the country which would now be deemed totally inadequate. . . . In a little vessel of scarcely over 200 tons, assisted by able and zealous officers under his command, many of whom were modelled under his hand and most of whom have since risen to eminence, he explored and surveyed the continent of South America. . . .

The Strait of Magellan, until then almost a sealed book, has since, mainly through his exertions, become a great highway for the commerce of the world—the path of countless ships of all nations; and the practical result to navigation of these severe and trying labours, which told deeply on the mental as well as the physical constitution of more than one engaged, is shown in the publication to the world of nearly a hundred charts bearing the names of FitzRoy and his officers, as well as the most admirably compiled directions for the guidance of the seamen which perhaps was ever written, and which has passed through five editions. . . .

His works . . . will be his most enduring monument, for they will be handed down to generations yet unborn.

Sir Roderick Murchison, the president of the Royal Geographical Society, who had presented the returning captain of the *Beagle* with the society's gold medal nearly thirty years before, told society members:

In deploring the loss of this eminent man who was truly esteemed by his former chief, the Prince of Naval Navigators, Sir Francis Beaufort, as by his successors, I may be allowed to suggest that if FitzRoy had not had thrown upon him the heavy and irritating responsibility of never being found at fault in any of his numerous forecasts of storms in our very changeful climate, his valuable life might have been preserved.

Apart from such appreciation, he was elsewhere quickly forgotten, remembered only, when remembered at all, for the name he gave to a most useful type of barometer, and by sailors for the rhymes in his *Weather Book*. He died in debt, his fortune exhausted. A collection was raised for Maria FitzRoy, to which Darwin contributed £100.

History—too often the reductive, diminishing view from the wrong end of a telescope—has known him, until recently, only as the facilitator, the lynchpin, for a far more famous man's revolutionary idea, an idea FitzRoy thought an abomination.

Yet 137 years after his death, FitzRoy's own work—his lonely, dogged, much ridiculed pursuit of an obsession with weather forecasting—was finally recognized by the British government's Meteorological Office (which he created). At noon on Monday, February 4, 2002, sea area Finisterre off the northwest coast of Spain—one of the twenty-four sea areas first established by FitzRoy for his telegraphed weather reports—was renamed FitzRoy. The BBC now broadcasts radio weather forecasts, vital for shipping, for these sea areas three times every twenty-four hours.

Mariners listen for his name daily.

ACKNOWLEDGMENTS

Many people have helped me with this book, and others, and in my writing life, in more ways than there is room to detail here. My gratitude and debt to them are too great to summarize in this finite space.

Sloan Harris, my friend and agent. Katharine Cluverius, Teri Steinberg at ICM. My editor at HarperCollins, Dan Conaway; and Cathy Hemming, Nikola Scott, Justin Loeber, Jane Passberger, Jason Curran, Martha Cameron, and Jill Schwartzman at HarperCollins. Andrew Franklin, Kate Griffin, and all at Profile Books, London. Anita and David Burdett; Anita particularly for her help at the PRO at Kew. Marion and Jeric Perth. Lucy Waitt at the National Maritime Museum, Greenwich, London; Helen Harrison of the State Library of New South Wales, Australia; Barbara Brownlie of the Alexander Turnbull Library, Wellington, New Zealand.

Sam Manning, again, for his beautiful maps, and Susan Manning for her crucial part in these.

At *Outside* magazine, Elizabeth Hightower, for more than she ever imagined, and Hal Espen. Elaine Lembo and Herb McCormick at *Cruising World*. Judith Miller and Beth Epstein at New York University in Paris.

Fraser Heston for the good word.

My mother, Barbara Nichols; David Nichols; Liz and Tony Sharp; Cynthia Hartshorn; Matt deGarmo; Penny and Robert Germaux; Peter Landesman; Doug Grant and Kathryn Van Dyke.

Josephine Franzheim, Kenneth Franzheim, Gay and Tony Brown for their spectacular hospitality and generosity.

The booksellers who over the years have invited me to their very special stores and championed my books—it would be unfair to single out one person or store without naming all; Booksense, and everyone associated with that wonderful enterprise; to all these people I am profoundly grateful.

Bennett Scheuer, a very true friend.

My wife, Roberta, for the faith, encouragement, and intelligence that allow me to reach higher, to write the right books.

SOURCES

My most constant guide was Robert FitzRoy, in his massive four-volume *Narrative of the Surveying Voyages of His Majesty's Ships* Adventure *and* Beagle *Between the Years 1826 and 1836 Describing Their Examination of the Southern Shores of South America and the* Beagle's *Circumnavigation of the Globe,* by FitzRoy, Captain Phillip Parker King, Captain Pringle Stokes, and, in Volume 3, Charles Darwin. Surely one of the most complete and exhaustive accounts of any (two) voyage(s) in history, this work is rich not only in its observations of wind, weather, sea and coast, nature ashore, and every possible aspect of a prolonged exploration of the globe by ship in the early nine-teenth century, but today it is most valuable for the view of the world of its main author, Captain Robert FitzRoy. He carried with him in all his doings the forthright prejudices of his age, and this gives his *Narrative* its great worth as a historical docu-ment. Only a few thousand copies were printed by its publisher Henry Colburn in 1839; Darwin's Volume 3, later retitled *Voy-age of the Beagle,* found a lasting readership and has never been out of print, but the full four-volume set is a rarity, going for $50,000 or more if one can be found. I was lucky to find a complete, virtually new, clothbound facsimile reprint, complete with charts, published by AMR Press of New York in 1966, for $350.

Janet Browne's magisterial two-volume biography of Charles Darwin, *Charles Darwin; Voyaging* (1995, Jonathan Cape, London) and *Charles Darwin; The Power of Place* (2002, Knopf, New York) was almost as useful. Browne's knowledge of Darwin and the world he lived in is encyclopedic, her authority absolute. Her book, with its bibliography, was a constant resource for me. Also very useful early on was the introduction Janet Browne wrote with Michael Neve to the Penguin edition of Darwin's *Voyage of the Beagle* (1989).

Of course, Charles Darwin himself. The 1839 (*Narrative, Vol. 3*, Colburn) and 1845 (John Murray, London; 1909, 1937, Collier/Harvard, New York) editions of his *Voyage*. His *Beagle Diary,* which formed the basis for his *Voyage*; I used the Cambridge University Press edition (paper, 2001, Cambridge), with useful additions, edited by Darwin's great-grandson, Richard Darwin Keynes.

Various editions of *Origin of Species*. Darwin's letters from the Folio Society edition of *A Narrative of the Voyage of H.M.S. Beagle*, edited by David Stanbury (1977, UK). *The Autobiography of Charles Darwin,* unabridged version edited by Nora Barlow (1958, Collins, London).

Many historic documents at the Public Records Office, Kew Gardens, London—ADM 1/1818; 1/2031; 51/3026; 51/3053; 52/3965; 53/239 among others—comprising the logbooks of H.M.S. *Beagle*; letters by Captains FitzRoy, Philip Parker King, and Pringle Stokes; particularly, most fascinatingly, King's long letter to his superiors at the Admiralty explaining the circumstances and details of the death of Stokes, the *Beagle*'s first commander (ADM 1/2031). To hold these documents in one's hand for an hour, to read these pre-Victorian captains' flourish-embellished handwriting, is worth months of factual research.

For most of the facts of Robert FitzRoy's life, *FitzRoy of the Beagle*, by H.E.L. Mellersh. London: Rupert Hart-Davis, 1968.

And then:

Ackroyd, Peter; *London*. London: Chatto & Windus, 2000.

Annan, Noel; *The Dons*. Chicago: University of Chicago Press, 1999.

Beaglehole, J.C.; *The Life of Captain James Cook*. London: A&C Black, 1974.

Bridges, E. Lucas. *Uttermost Part of the Earth*. London: Hodder & Stoughton, 1951.

Caldwell, John. *Desperate Voyage*. London: Victor Gollancz, 1952.

Dugan, James. *The Great Iron Ship*. New York: Harper Brothers, 1953.

Fowles, John. *The French Lieutenant's Woman*. New York: Little, Brown, 1969.

Freeman, Michael. *Railways and the Victorian Imagination*. New Haven: Yale University Press, 1999.

Gosse, Edmund. *Father and Son*. London: Penguin, 1986.

Gosse, Philip Henry. *Omphalos*. London: John Van Voorst, 1857.

Hazlewood, Nick. *Savage*. London: Hodder & Stoughton, 2001.

Hayter, Alethea. *A Sultry Month*. London: Robin Clark, 1992.

Hiscock, Eric. *Cruising Under Sail*. Oxford: Oxford University Press, 1950.

Hughes-Hallet, Penelope. *The Immortal Dinner*. London: Viking, 2000.

Hydrographer of the Navy. *South America Pilot*, Volume 2. United Kingdom: Hydrographic Office of the British Royal Navy, 1993.

Inwood, Stephen. *A History of London*. New York: Carroll & Graf, 1998.

Marquardt, Karl Heinz. *H.M.S.* Beagle. London: Conway Maritime Press, 1997.

Morris, James. *Heaven's Command; Pax Britannica*. New York: Harcourt Brace, 1973.

Moorehead, Alan. *Darwin and the* Beagle. London: Hamish Hamilton, 1969.

Pakenham, Thomas. *The Scramble for Africa*. London: Abacus, 1992.

Reisenberg, Felix. *Cape Horn*. New York: Dodd, Mead, 1939.

Rich, Louise Dickenson. *The Coast of Maine*. Maine: Down East Books, 1993.

Ritchie, Rear-Admiral G.S. *The Admiralty Chart.* London: Pentland Press, 1995.

Robinson, W.A. *10,000 Leagues Over the Sea.* New York: Harcourt Brace, 1932.

Rodger, N.A.M. *The Wooden World.* New York: Norton, 1996.

Severin, Tim. *The Spice Islands Voyage.* London: Little, Brown, 1997.

Slocum, Joshua. *Sailing Alone Around the World.* New York: The Century Company, 1900.

Thwaite, Ann. *Glimpses of the Wonderful; The Life of Philip Henry Gosse.* London: Faber & Faber, 2002.

Vaughan, Adrian. *Brunel.* London: John Murray, 1993.

Wallace, Alfred Russell. *The Malay Archipelago.* London: Macmillan, 1869; *My Life.* London: Chapman and Hall, 1905.

Zimmer, Carl. *Evolution.* New York: HarperCollins, 2001.